T0136427

About Island Press

Since 1984, the nonprofit organization Island Press has been stimulating, shaping, and communicating ideas that are essential for solving environmental problems worldwide. With more than 1,000 titles in print and some 30 new releases each year, we are the nation's leading publisher on environmental issues. We identify innovative thinkers and emerging trends in the environmental field. We work with world-renowned experts and authors to develop cross-disciplinary solutions to environmental challenges.

Island Press designs and executes educational campaigns, in conjunction with our authors, to communicate their critical messages in print, in person, and online using the latest technologies, innovative programs, and the media. Our goal is to reach targeted audiences—scientists, policy makers, environmental advocates, urban planners, the media, and concerned citizens—with information that can be used to create the framework for long-term ecological health and human well-being.

Island Press gratefully acknowledges major support from The Bobolink Foundation, Caldera Foundation, The Curtis and Edith Munson Foundation, The Forrest C. and Frances H. Lattner Foundation, The JPB Foundation, The Kresge Foundation, The Summit Charitable Foundation, Inc., and many other generous organizations and individuals.

The opinions expressed in this book are those of the author(s) and do not necessarily reflect the views of our supporters.

Hazardous Seas

Hazardous Seas

A Sociotechnical Framework for Early Tsunami Detection and Warning

Edited by Louise K. Comfort and Harkunti P. Rahayu

ISLANDPRESS | Washington | Covelo

Library of Congress Control Number: 2022946268

All Island Press books are printed on environmentally responsible materials.

Manufactured in the United States of America
10 9 8 7 6 5 4 3 2 1

Keywords: atypical tsunami; Badan Pengkajian dan Penerapan Teknologi; Badan Riset dan Inovasi Nasional; bottom pressure sensors; bowtie architecture for community resilience; BPPT; BRIN; cable-acoustic hybrid tsunami early warning system; CBT Labuan Bajo; community resilience framework; community tsunami preparedness; community-based tsunami shelters; complex adaptive systems; device-to-device communication; disaster risk reduction; DRR; end-to-end tsunami warning and mitigation system; horizontal acoustic communication system; InaCBT— Indonesian Cable-Based Tsunameter; InaTEWS; Indian Ocean Tsunami Warning and Mitigation System; Indian Ocean wave exercise; Indonesian-American cooperation for tsunami early warning; Indonesia Tsunami Early Warning System; Institut Teknologi Bandung; IOTWMS; IOWAVE; ITB; mainstreaming tsunami DRR; Mentawai Tsunami Early Warning System; multimodal downstream warning chain; near-field tsunami; ocean bottom unit; Padang Tsunami Early Warning System; people-centered early warning system; scalable wireless network for disaster; shelter evacuation system; shelter selection, design, and construction; single-board computer-based network; smartphone tsunami applications; system dynamics evacuation model; system prototype deployment; tsunami preparedness; undersea acoustic tsunami detection network; undersea research station; United Nations Decade of Ocean Science; United Nations Tsunami Program Framework; upstream and downstream warning chain

CONTENTS

Chapter 8. Real-Time Seafloor Tsunami Detection and
Acoustic Communications   199
Lee Freitag, Keenan Ball, Peter Koski, James Partan, Sandipa Singh,
Dennis Giaya, and Kayleah Griffen

Chapter 9. A Prototype Ocean Bottom Pressure Sensor Deployed
in the Mentawai Channel, Central Sumatra, Indonesia:
Preliminary Results   221
Emile A. Okal and Lee Freitag

Chapter 10. Underwater Sensor Network Prototype for Tsunami
Detection and Warning: A Long Deployment Journey toward
Functionality   247
X. Xerandy, Iyan Turyana, Lee Freitag, Wahyu W. Pandoe,
Harkunti P. Rahayu, and Febrin Anas Ismail

Chapter 11. Indian Ocean Tsunami Warning and Mitigation
System: Initiation, Evolution, and Implementation   275
Harkunti P. Rahayu

Chapter 12. Creating a Sustainable Learning System in Regions
of Risk   291
Louise K. Comfort, Wahyu W. Pandoe, Harkunti P. Rahayu, and
Iyan Turyana

Afterword   313

A SCIENTIFIC NOTE

As Indonesia and the world face the continuing consequences of climate change and rapid increase in extreme hazards, it is imperative that we develop and expand the knowledge of science, technology, and social action to mitigate the risk of devastating tsunamis to our coastal communities. The chapters of this book, *Hazardous Seas*, present a proven strategy to reduce the threat of deadly tsunamis and engage coastal communities in building resilient practices to protect lives and livelihoods. Importantly, the design of the prototype early tsunami detection and warning system combines the science of the ocean environment with the technologies of ocean engineering to validate the function of undersea acoustic communication. The prototype further includes the role of social networks supported by wireless communication technologies to inform and engage community residents in self-organizing action to reduce risk from tsunamis.

The prototype harnesses the science of the warm equatorial waters of the Mentawai Channel in Indonesia to extend the range of undersea acoustic communication to 25 to 30 kilometers and uses engineering technologies to design, build, implement, and test an undersea network in the deep waters of the Mentawai Sea in practice. The network is designed to detect the subsurface movement of water that indicates a tsunami wave and to communicate data directly to the Indonesian Agency for Meteorology, Climatology, and Geophysics (BMKG) via satellite within minutes, confirming an incoming tsunami wave. When the system is ready to be implemented in the existing operational system of the Indonesian Tsunami Early Warning System (InaTEWS), it will be a critical addition to InaTEWS, representing a major advance in current tsunami mitigation and warning practice for Indonesia.

The prototype also focuses on the role of social action in communities exposed to tsunami hazards. Included in the prototype is the design, implementation, and testing of an electronic network for communication in a disaster-damaged neighborhood. This electronic network supports local social networks, with neighbors being trained to communicate with one another and follow the quickest routes to safety. Both the undersea network and the community network strengthen the InaTEWS at its most vulnerable points: the initial detection and confirmation of a tsunami wave and the final communication of that risk to neighborhood residents that enables them to reach safe shelter.

Although the findings from the development, implementation, and testing of this prototype early detection and warning system will benefit Indonesia, the other twenty-seven member states of the Indian Ocean Tsunami Warning and Mitigation System will benefit as well. The systematic inquiry presented here offers a bold contribution to the development of innovative, cost-effective programs to protect coastal communities of the Indian Ocean Basin from the destructive potential of sudden and severe near-field tsunamis and other coastal hazards. Although the findings documented in this book constitute but a first step in the continuing international and interdisciplinary inquiry to reduce the catastrophic risk of near-field tsunamis, they represent a major contribution to science, ocean engineering, community education, and resilience to hazards consistent with the goals of the United Nations Decade for Ocean Science and Sustainable Development.

Prof. Dwikorita Karnawati, PhD
Director, Indonesian Agency for Meteorology, Climatology, and
    Geophysics (BMKG)
Chair, UNESCO Intergovernmental Oceanographic Commission/
    Intergovernmental Coordination Group, Indian Ocean Tsunami
    Warning and Mitigation System
Jakarta, Indonesia

FOREWORD

The gigantic Aceh tsunami in 2004 and other tsunamis that have followed—up until now—always bring along impromptu surprises and self-shocking elements, whether or not an event claimed any victims. There are never two exactly similar tsunamis, either by the location of epicenter or the number of fatalities and amount of economic losses, but they always pose apprehension and anxiety.

Casualties and losses due to tsunamis are often associated with both structural and nonstructural problems, which are further manifested in the need for state-of-the-art tsunami early warning technology, awareness, and literacy of the impacted community. Learning from the progression of the Early Warning System of the United Nations International Strategy for Disaster Reduction (UN-ISDR) and the United Nations Development Programme (UNDP), initiated in 2004 and advanced in 2009, the 2019 Indian Ocean Tsunami Warning and Mitigation System (IOTWMS) Assembly summarized the thoughts into three strategic pillars: (1) hazard, risk assessment, and reduction; (2) rapid detection, warning, and dissemination; and (3) public education, evacuation, emergency planning, and response. In essence, it underlines the importance of an end-to-end approach, integrated from upstream to downstream, which necessitates understanding the sources of tsunamis, prompt detection mechanisms, and swift relay of information that reaches out to communities.

After the 2004 Aceh tsunami, Indonesia, through international support and cooperation among national government institutions, established the tectonic-based Indonesia Tsunami Early Warning System (InaTEWS). It became part of the regional tsunami warning provider for twenty-eight countries surrounding the Indian Ocean, as well as Australia and India, and later "examined" various types and levels of tsunami events. For tectonic events, the system has proven to disseminate information for potential tsunamis in a fast and accurate manner and reach out to wide-ranging locations. Amid the increasingly varied causes of tsunamis—for example, those that occurred in Indonesia's Central Sulawesi, Sunda Strait, and Central Maluku in 2018 and 2020—national experts and international communities recognized that InaTEWS is no longer sufficient.

The government of Indonesia is very aware of the need to intensify and encourage mitigation efforts and improve the existing InaTEWS to face the various causes of tsunami events. Thus, through Presidential Decree No. 93, Year of 2019 regarding strengthening and developing information on earthquake and tsunami early warnings, the Indonesian government research institute, the Agency for the Assessment and Application of Technology (BPPT), has been tasked to develop, operate, and conduct maintenance of a tsunami wave detection system. The assessment and application of InaBUOY, InaCBT (cable-based tsunameter), and InaCAT (coastal acoustic tomography) and their deployment at several locations in Indonesian waters, among other places, exemplify the efforts in which all systems are integrated with the InaTEWS at the Indonesian Agency for Meteorology, Climatology and Geophysics (BMKG). The establishment of a tsunami observation center (InaTOC) at BPPT serves as an intermediate system to guarantee that all data and information are connected and transmitted.

Since 2013, BPPT has also collaborated with international researchers to improve national capacity in tsunami disaster mitigation. BPPT, the Bandung Institute of Technology (ITB), and Andalas University have been working with the University of Pittsburgh and the Woods Hole Oceanographic Institution in the United States in a research project that explores an interdisciplinary and sociotechnical approach to increase community resilience to tsunami. This research project develops a tsunami detection system that combines underwater fiber-optic cable, underwater sensors, and acoustic technology. The prototype is located near Siberut Island in West Sumatra, Indonesia. In this project, BPPT has contributed to the system's deployment effort by providing its research vessel *Baruna Jaya IV* and ocean bottom unit canister.

The Intergovernmental Oceanographic Commission–United Nations Educational, Scientific and Cultural Organization (IOC–UNESCO) 2019 Working Group on Tsunami and Other Ocean Warning Systems stated that despite the increasingly advanced development of tsunami detection technology—tsunami radar, buoys, global navigation satellite systems, infrasound, smart cables, and so forth—the main challenge facing tsunami warning systems lies in how to reduce the level of uncertainty of tsunami detection so that potentially affected coastal communities may receive a higher degree of certainty to start evacuation as soon as possible.

The vision of a safe ocean within the framework of the UN Decade of Ocean Science states that the overall objective for the assessment and application of tsunami mitigation technology, whatever form of the contribution, is to equip those who live, work, and recreate in the potentially affected coastal areas with the knowledge and provision to act before the first tsunami wave strikes. The threat of disaster thus demands that tsunamis be understood and observed so that their impact can be predicted accurately.

Hazardous Seas: A Sociotechnical Framework for Early Tsunami Detection and Warning presents a very meaningful contribution to build a safe ocean society and tsunami-ready community. I convey my high appreciation to all the coauthors of this book for their engagement. Each chapter manifests itself as an intellectual contribution to this effort, not only for Indonesia, but for worldwide coastal communities, today and beyond.

Prof. Dr. Ir. Hammam Riza, MSc, IPU
Professor of Artificial Intelligence, Electrical and Computer
    Engineering
University of Syiah Kuala (USK), Aceh, Indonesia
Chairman, Agency for Assessment and Application of Technology
    (2019–2021)
Jakarta, Indonesia

At a tea break during a workshop in Hawaii in February 2006, the concept for this project emerged in a conversation with Dr. Idwan Suhardi and several of his colleagues from the Ministry of Research and Technology of Indonesia. The workshop, organized by three colleagues from the University of Pittsburgh—Daniel Mosse, Louise Comfort, and Taieb Znati—focused on exploratory research on sensor-based infrastructure for early tsunami detection. With support from the US National Science Foundation (NSF), we invited a small group of experts from Asian nations affected by the 2004 Indian Ocean Tsunami to consider possible, but practical, strategies for early detection and warning of tsunamis. Suhardi, mindful of his nation's painful loss of more than 126,000 lives in the December 26, 2004, tsunami, asked if there were new technologies available that could assist Indonesia in planning for tsunami mitigation, both in early detection of an oncoming tsunami and in timely warning for a threatened coastal community. The organizers of the workshop offered to explore the possibility of forming an international and interdisciplinary team to design innovative methods for early detection and warning for communities at risk from tsunamis.

The conversations, insights, and reports shared by the international workshop participants led to a second NSF grant funded under the Decision, Risk, and Uncertainty (DRU) Program, *Designing Resilience for Communities at Risk: Improving Decision Making to Support Collective Action under Stress*, beginning on September 1, 2007. This project explored the intersection of technical and organizational systems to develop a model for a low-cost, energy-efficient, early tsunami detection and warning system in conjunction with Indonesian colleagues at the Bandung Institute of Technology (ITB), Bandung, and Andalas University, Padang, Indonesia. The project also received letters of support from three major Indonesian governmental agencies: Ministry of Research and Technology (RISTEK); Ministry of Meteorology, Climatology, and Geophysics (BMKG); and the Agency for the Assessment and Application of Technology (BPPT). Visits to Indonesia by US researchers forged strong bonds with Indonesian researchers and government personnel to create an international and interdisciplinary research team and provided an invaluable opportunity to see and understand the local conditions of tsunami risk for West Sumatra. As the project closed in 2012, the research team had produced a set of computational models for both an undersea network for early tsunami detection and a community-based wireless network to support communication among neighborhood residents during evacuation in an actual tsunami event.

With vital collaboration and strong support from Indonesian colleagues, the models developed in the DRU project served as the basis for a third grant from the NSF, Hazard SEES Type 2, OCE 1331463, beginning in September 2013. The project proposed to design, build, and test a prototype early tsunami detection and warning system in an actual risk environment. The project involved two interconnected networks that created a sociotechnical system in practice. First, the prototype undersea network would detect a tsunami wave and transmit the warning via acoustic communication to a shore station that would relay the data via satellite to scientists at BMKG. The BMKG scientists would confirm the threat of tsunami and transmit the warning through the Indonesian Tsunami Early Warning System (InaTEWS) to the local emergency operations centers (EOCs) for the threatened community. Second, personnel at the local EOCs, provincial and municipal, would transmit the warning to local neighborhood networks to alert the residents to evacuate. This alert would activate a community-based wireless network at the neighborhood level to aid communication among residents in rapid evacuation of areas at risk. The two networks, operating in coordination, would strengthen the most vulnerable points of the InaTEWS, the initial detection of a tsunami wave and the last link of communication to neighborhood residents at risk, increasing the timeliness and accuracy of the warning to coastal communities.

At its inception, this international and interdisciplinary project included researchers from six institutions: University of Pittsburgh (Pitt), Carnegie Mellon University, Northwestern University, and Woods Hole Oceanographic Institution (WHOI) from the United States and the Bandung Institute of Technology and Andalas University from Indonesia. Padang City in West Sumatra served as the field site for the project and location of the neighborhood network development, while the Central Channel of the Mentawai Sea near Siberut Island served as the field site for the undersea network. See figure P-1.

Initiating the Hazard SEES (Interdisciplinary Research in Hazards and Disasters) project was the easy task; designing, building, implementing, and testing the prototype networks in actual field environments half

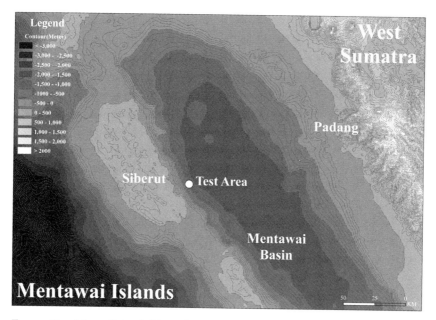

Legend
Contour(Meter)
< -3,000
-3,000 - -2,500
-2,500 - -2,000
-2,000 - -1,500
-1,500 - -1,000
-1,000 - -500
-500 - 0
0 - 500
500 - 1,000
1,000 - 1,500
1,500 - 2,000
> 2000

West Sumatra

Padang

Siberut • **Test Area**

Mentawai Basin

Mentawai Islands

50   25   0 KM

FIGURE P-1. Map of the field study area: Central Channel, Mentawai Sea, and West Sumatra, Indonesia.

a world away was the hard part. The project was designed to be completed within four years, 2013–2017, with active participation from Indonesian academic colleagues and strong support from Indonesian agencies. Working in an actual environment with multiple conflicting pressures presented multiple challenges. First, Indonesia elected a new president in 2014. The previous president, who had been in office during the 2004 tsunami, and his officials in major ministries had been very supportive of the project, but the new administration had different priorities and selected different candidates to lead the ministries, requiring a new round of presentations, demonstrations, and discussions to earn their support.

Second, an economic downturn in Indonesia in 2014–2015 led to a 40 percent budget cut for the Indonesian ministries, seriously reducing their capacity to contribute time and resources to the project as initially planned and delaying the implementation of the undersea prototype. Nonetheless, work on the neighborhood network proceeded in 2015 and 2016. A proof of concept for the undersea acoustic network was successfully completed in 2016, but delays in schedules and funding pushed the actual implementation of the undersea network back a year. The project requested and received a no-cost extension from the NSF to complete the project, originally planned for 2017, to the following year. As plans and resources were gradually restored for full deployment of the undersea network in late August 2018, coordinating the schedules among the three Indonesian ministries that were vital to support the prototype early warning again proved unworkable, and plans for implementation of the undersea prototype that year were again canceled.

On September 19, 2018, nature intervened when the Palu tsunami occurred on Sulawesi, killing more than four thousand people, injuring more than ten thousand, and leading to the evacuation of more than seventy thousand. This sobering event underscored the importance of developing innovative methods of early tsunami detection and warning. At this point, the continuation of the project was at risk as the NSF funding for the project ended on August 30, 2018. The global community had watched both the tragedy of the Palu event and the near completion of the prototype early tsunami detection system, though. Noting the potential for the undersea early tsunami detection prototype to protect lives and communities in tsunami-prone areas, the Swiss Re Foundation, a global foundation committed to risk reduction, contacted Comfort to inquire about the requirements to complete the deployment and testing of the prototype early tsunami detection system and invited submission of a proposal to fund the completion of the planned deployment in Indonesia.

With funding from the Swiss Re Foundation confirmed, the international interdisciplinary team, now smaller as the task focused primarily on the implementation of the undersea network, reviewed and reorganized plans for deployment of the prototype system in the Mentawai Sea. A formal agreement among BPPT, WHOI, and Pitt was signed to supplement the existing partnership between Pitt and ITB in November 2019 to conduct the deployment of the undersea network in late December 2019. With minor adjustments to the schedule, the deployment was conducted from January 1 to January 3, 2020, in the Mentawai Sea. This deployment produced the successful outcome of validating the undersea network communication for the prototype detection system, but a failed deployment of the cable connecting the acoustic modem to the shore station. To correct this key link in the network, it was essential to lay a new cable and establish a working connection between the undersea detection and acoustic transmission instruments and the onshore relay station.

The international network had forged a strong commitment to complete the full deployment of the prototype. The Swiss Re Foundation agreed to fund a second grant to purchase a new cable. BPPT offered to contribute the ship time and crew to make a second cruise to the

Siberut site to deploy the cable. The WHOI engineers worked with the Indonesian engineers to review and redesign the necessary equipment to complete a second deployment, which would require laying new cable and refabricating the equipment needed for a fresh connection. All plans were in place, and the cruise to Siberut to install the new cable was scheduled for late March 2020. But external events intruded again, this time in the form of a global pandemic, COVID-19, that shut down international travel and closed the engineering labs of BPPT. The March 2020 cruise was delayed again.

Recognizing the importance of reestablishing the connection between the shore station and the undersea ocean bottom unit (OBU) that housed the equipment that had been deployed successfully in early January 2020, the Swiss Re Foundation offered a small grant to meet the additional costs of continuing the project under COVID-19 restrictions. With full support from BPPT in terms of providing ship and crew for the cruise, a new cable funded by the Swiss Re Foundation, and a professional company hired to lay the cable, the Indonesian engineering team undertook a second deployment in July 2020. Regrettably, travel restrictions imposed to limit exposure to COVID-19 prevented the WHOI engineers from joining the cruise for the second deployment. With careful planning and simulation of the seabed route, the cable was laid professionally. Improvements were also made to the shore station, with solar installation to ensure electrical power and tests performed to ensure an effective satellite connection for data transmission. The final test, however, revealed that the canister had failed under the pressure of water from a depth of more than 700 meters. Replacing the canister at substantial expense was the only solution.

With the dedication and commitment of true innovators, the international interdisciplinary research team conducted a thorough review of the requirements needed for a final successful deployment of the prototype early tsunami detection and warning system. The science had already been proven. Acoustic communication underwater for at least 25 kilometers had been documented twice. The challenge was implementing the engineering required for the system to operate in deep water. After careful consideration, BPPT agreed to fund the fabrication of a new canister to meet international standards. The Swiss Re Foundation agreed to the reallocation of monies that had been designated for an international conference to publicize the findings from the prototype system to expenses for a third deployment. The WHOI engineers continued to collaborate long-distance with the Indonesian engineers to review and support the plans for a third deployment, initially planned for early August but delayed again until early September by a major surge of COVID-19 in Indonesia.

On September 1, 2021, the project experienced a further external shock when the implementing agency, BPPT, was dissolved and a new agency, the National Research and Innovation Agency (BRIN), absorbed most of the former BPPT responsibilities. This organizational change created a period of reorganization and redefinition of responsibilities as staff who had played primary roles in the project were now shifted to new units and new positions. Still, the Indonesian engineers and researchers were committed to the project and continued to adapt to the new administrative arrangements.

The challenge for the third deployment was that only the canister needed to be replaced, as the cable had been successfully laid. It would be necessary to retrieve the old canister at the bottom of the sea and replace it with the new canister at sea. This exchange would require a ship with a dynamic positioning system to hold the ship steady in rolling ocean waters while the engineers performed the delicate task of disconnecting the old canister and reconnecting a new canister. BPPT did not have such a ship in its fleet, which meant that it would be necessary to hire such a ship with its crew, at substantial cost. The monies remaining from the Swiss Re Foundation grant were insufficient to cover the cost. Pitt and ITB personnel agreed to search for funding to support a third, and final, deployment.

After the tragedy of the Palu tsunami in 2018, officers of GOJEK, the Indonesian ride-hailing company, had contacted Comfort, principal investigator of the Hazard SEES project, to offer its assistance in completing the deployment of the prototype early tsunami detection and warning system. Recalling that offer, Comfort contacted GOJEK and explained the situation of the prototype and the need for additional funding to complete it successfully. The request was referred to the Yayasan Asak Bangsa Bisa (YABB) Foundation, newly established by GOJEK to serve humanitarian needs in Indonesia and four other countries in Southeast Asia.

YABB staff responded with a request for more information regarding the project and, after review and discussion, invited a proposal for funding. YABB agreed to fund the proposal, but initially only at half the amount requested. As the negotiations for a ship with dynamic positioning continued and the detailed costs for the third deployment became clear, the proposal was revised to increase the request for funds that would cover the additional cost of a ship with DP Ship Management and related expenses. After detailed negotiations, YABB agreed to increase its contribution to cover the cost of the DP ship and crew.

Timing was critical in this process, and the ship was available only for a specific window early December 2021. The WHOI engineers had prepared to join the crew for the cruise to Siberut, filing applications for travel visas, purchasing tickets, and planning to leave in late November. As they waited for confirmation of travel visas, external events intruded again as the Omicron variant of the SARSCoV2 virus surged across the world in late November 2021. The Indonesian Office of Immigration again closed the country's borders to foreign travelers to protect the population against infection, and the WHOI engineers were not granted visas for travel.

All other plans for the redeployment of the canister were in place. The canister had been fabricated according to international standards and tested successfully, and the DP ship and crew were headed toward Siberut. The international engineering team proceeded with the deployment of the canister, the most difficult and risky task in activating the undersea network. In a remarkable instance of "tele-engineering," the WHOI engineers were in continual communication with BPPT/BRIN engineers via Zoom and WhatsApp across a 12-hour time difference as the team succeeded in replacing the canister and establishing a successful connection to the shore station.

Given events beyond their control, the new BRIN administrators extended the funds for a fourth cruise to Siberut to the end of March 2022. This cruise would retrieve the OBU for renovation and reconnect it to the now-working canister and cabled modem. In March 2022, the Omicron variant receded; visas were available in early April, and the WHOI engineers traveled to Indonesia in mid-April 2022. They met the Indonesian team in Padang and traveled to Siberut, where they were able to retrieve the OBU, renovate the equipment, replace the batteries, test all connections, and successfully redeploy the OBU. The final step was to reactivate the cabled modem that had been tested successfully in December 2021. Apparently, between December 2021 and April 2022, an electrical short had occurred in the cable line under deep water. All components of the prototype system had been tested and proven to work at different times, and the science of underwater acoustic communication over long distances had been demonstrated three times; still, the major engineering task of integrating all components in deep ocean waters requires specialized equipment and investment. The strain of interrupting the deployment at three different stages due to external conditions presented an unexpected challenge in the effort to activate a fully working undersea prototype early tsunami detection and warning system.

Looking back over the sixteen years since an early concept for mitigating tsunami risk was articulated over tea at a workshop in Hawaii, we confirm that this vision has been translated into reality. The challenges were many. It was not easy to secure sufficient funding to support the technical development. The task of integrating a shared vision among multiple organizations necessary to complete this task, given competing interests, changing priorities, and varying schedules, was equally challenging. The engineering and computational designs involved in both the neighborhood wireless network and the undersea acoustic network were implemented and tested in physical environments that had never been tried before. Nevertheless, a small group of dedicated researchers, decision makers, and practicing engineers continued to overcome obstacles and engage supporters from a wide range of governmental, academic, and nonprofit organizations to achieve a working prototype early tsunami detection system. The work is not done—there are still many steps left to achieve full implementation of the basic design and promise of the system—but the vision of early tsunami detection and warning is clearer and brighter now, with demonstrated evidence that such a sociotechnical system is workable in practice.

Louise Comfort
Oakland, California, USA

Harkunti Rahayu
Bandung, Indonesia

Chapter 1

Building Community Resilience to Disaster Risk: A Sociotechnical Approach

Louise K. Comfort and Mark W. Dunn

Communities exposed to intermittent but recurring risk confront the difficult problem of recognizing when risk is imminent and mobilizing collective action to reduce harm. In many instances, signals are scattered, timing is uncertain, and alternatives for action are not clear; yet the costs and consequences of inaction in areas of known risk are enormous in terms of loss of life, damage to property, and disruption of economic and social activities when hazards strike.

Recent tsunamis in Indonesia give deadly evidence of the challenge confronting public policy makers who are charged with legal responsibility to protect their citizens from harm. On September 29, 2018, more than two thousand lives were lost in Palu, Sulawesi, when a tsunami wave, or run-up, more than 5 meters high crashed ashore and exacerbated severe liquefaction that ensued from the precipitating earthquake, destroying most of an adjacent town (USGS 2018; Wright 2018). Less than three months later, on December 23, 2018, tsunami waves more than 3 meters high, generated by undersea landslides from the erupting volcano, Anak Krakatau in the Sunda Strait, damaged towns on both Java and Sumatra islands and claimed at least 373 lives (Martin and Zhou 2018). These events pale in comparison to the severe losses incurred from the December 26, 2004, Indian Ocean tsunami that took an estimated 230,000 lives, more than 126,000 in Indonesia. These sudden, massive, deadly events require the active engagement of the whole community to anticipate risk and reduce harm.

How does a community learn to recognize risk and act collectively to reduce probable loss from extreme events? This extraordinarily complex, interdisciplinary, interorganizational, and interjurisdictional problem requires knowledge of the science underlying the hazard; knowledge of the

1

laws, policies, and plans for mobilizing public services to protect community residents; knowledge of both resources and risks in the community in terms of infrastructure, economic situations, and social organizations; and importantly, knowledge of the local culture, beliefs, values, and customs of the residents. No single organization, agency, or jurisdiction can manage to recognize these hazards and reduce risk to the exposed communities alone.

Decision making in risk environments has a long, rich tradition of study in public policy and administration research and theory (La Porte 1975; Beck 1992; Roberts 1993; Weick 1993). This tradition acknowledges the uncertainty involved in risk environments, but largely views organization as a means of establishing sufficiently robust policies and procedures to maintain reliable performance in changing conditions (Roberts 1993; Weick 1995; Weick and Sutcliffe 2007) and adequate resources to meet unexpected demands (Landau 1969; Van de Walle 2014). In extreme environments, organizations designed to maintain order and stability tend to fracture under pressure as their structures of authority no longer fit the altered operating conditions (Weick 1993; Comfort 2019). Other researchers have focused on the role of communication as a key factor in integrating disparate views, personnel, and cognitive assessments of changing conditions (Luhmann 1996; Comfort 2007; Castells 2009). Communication, central to creating the capacity for collective cognition, enables informed action on a community-wide scale and builds resilience to recurring hazards. The actual development of cognitive capacity in practice is not well understood, however, nor is the crucial transition from cognition to action on a community-wide scale. Lack of understanding of these related cognitive processes represents a critical gap in the wider literature on managing risk in complex environments.

Collective Cognition in Complex Adaptive Systems

This book focuses on the problem of collective cognition to enable a community to mobilize quickly to reduce harm. The following chapters address collective cognition as a series of iterative learning experiences among multiple organizations, using the framework of complex adaptive systems of systems (CASoS) (Glass et al. 2011). The CASoS framework, developed by interdisciplinary research teams at Sandia National Laboratories, addresses large-scale eco-socioeconomic-technical systems by adapting their physical, engineered, social, and economic systems to reduce that risk. These systems include the interactions between recurring hazards that threaten communities and the communities' capacity to respond to known hazards. The CASoS analytical framework includes four basic research tasks (Glass et al. 2011, 882–83):

1. Define the policy problem and set it in a theoretical context of complex adaptive systems.
2. Design a conceptual computational model for addressing the policy problem.
3. Implement and test the computational model in an actual environment of risk.
4. Evaluate the performance of the test model and make recommendations for policy and practice.

The learning process evolves continuously as interactions among subsystems adapt to one another and change the environment in which they are operating, precipitating fresh adaptations at different levels of performance in the larger metasystem. The process functions through the search and exchange of information among the component systems; the extent to which the metasystem, or system of systems, adapts quickly and effectively to sudden, emergent threats depends on the technical and organizational infrastructures for communication that are available to the residents and actors in the threatened community. This set of interactions creates the basis for collective cognition of risk, or shared understanding of an imminent threat that informs collective action (Weick 1993; Argyris and Schön 1996).

Designing a learning process to manage urgent events like near-field tsunami risk is fundamentally a sociotechnical problem. In this book, we present the design, process, models, and early results of a set of integrated electronic networks to address the threat of near-field tsunamis to the city of Padang, West Sumatra, Indonesia. In the prototype system developed under this project, the initial threat of a tsunami is detected by an undersea sensor that transmits data acoustically to a cabled receiver and shore station. The aggregated data are then transmitted via satellite to the Indonesian scientific agency BMKG (Agency for Meteorology, Climate, and Geophysics) in Jakarta for validation and issuance of a warning to the community at risk. This process occurs in minutes. The warning is then communicated via satellite to the organizational network of emergency response agencies to alert the residents of the community at risk. An additional prototype of a distributed capability for community warning has been recently demonstrated with handheld electronic devices connected through a wireless network that operates in a disaster-degraded environment. The entire process must occur within 20 to 30 minutes, before the tsunami wave strikes shore, to provide effective warning.

Two points in this process are especially challenging. The first is the initial detection and confirmation of a tsunami wave that would threaten the community. Our research presents the design, field tests, deployment, and first results for an undersea network to detect and communicate the signals of a tsunami wave, using underwater acoustic communications. The second point is the communication of the tsunami warning to residents at the neighborhood level so that all members of the community receive the alert and can take informed action. In many past tsunami events, it is the failure of this "last mile" of communication to neighborhood residents that has resulted in sobering loss of life. Our research presents the design, models, and simulation results of a network of electronic devices for multiway communication to guide neighborhood residents to safe shelter. The two networks, undersea and local neighborhood, together strengthen the most problematic components of the existing Indonesian Tsunami Early Warning System (InaTEWS) to create an integrated flow of information across technical, organizational, and neighborhood channels that enable a community to recognize and respond to tsunami risk in a timely, informed way. In practice, these innovative networks increase the potential of InaTEWS to save lives and reduce losses from sudden, devastating tsunamis.

Individual Learning in Risk Environments

Seeking to understand how a community develops collective cognitive capacity to recognize risk in practice, we review the constraints on learning processes in urgent contexts. Cognitive capacity is defined, first, in individual terms, as the amount of information an individual can hold in working memory and use for immediate problem solving (Rittel and Webber 1973). With limited cognitive capacity, individuals are constrained in managing complex, rapidly changing, urgent environments (Simon 1996; Kahneman 2012). Individuals function in two learning modes, short-term problem solving capacity, and long-term recall. Daniel Kahneman (2012) characterized these modes as system 1, or fast learning that relies on short-term memory, but is prone to error, and system 2, or slow learning, that relies on long-term memory, recall, and corrective capacity, but costs time, attention, and effort. Operating under ordinary conditions, the two modes of learning complement each other, with system 1 allowing rapid decision making under tight time constraints and system 2 bringing the perspective of long-term memory to correct hastily made assumptions and rushed judgment based on incomplete observations. Under emergency conditions, interaction between the two modes of learning escalates, and the potential for cognitive overload of both learning systems increases, leading to cascading error.

Emergency and disaster response environments are fast-paced, high-stress settings with massive, diverse information flows in the form of messages, visual cues, signals, and noise. Initial information in disaster environments is often incomplete, ambiguous, and changing. Personnel operating in these environments confront critical incoming details while attempting to select, organize, and integrate relevant information into a current operational picture. Given limited cognitive capacity, individuals struggle in such environments with high demands on their attention for information processing (Rittel and Webber 1973). When information flow increases, problem-solving capacity drops as individuals shut out new information or sources of knowledge to focus on immediate tasks (Klein et al. 1993). In effect, the volume of information flow exceeds the individual's cognitive capacity, or amount of information that the individual can hold in working memory at any one time (Sweller 1988). To carry out basic cognitive functions in this demanding environment, a person necessarily sheds distractions or reduces external dissonance.

Importantly, the capacity of operations personnel as well as community residents to absorb new information and translate it into action is related to their previous level of knowledge and experience (Cohen and Levinthal 1990; Klein et al. 1993). This elasticity creates the possibility of redesigning the channels of information flow within the community and raising the level of awareness and understanding of known risk among individual members. This nexus of information and action is critical to building the capacity of the community to function as a coherent system of systems in managing intermittent but recurring risk.

The primary tasks for improving decision making in urgent, rapidly changing environments are fourfold: (1) minimize cognitive load for individual decision makers; (2) increase their level of awareness of threats in the immediate situation; (3) represent timely, valid information clearly; and (4) enable rapid communication and commitment to action as decision makers discover viable strategies for reducing risk emerging in the broader community (Paas and Sweller 2014). In complex environments, the design and use of information technologies to activate learning processes on multiple channels simultaneously engages diverse audiences in assessing imminent risk and updating strategies of informed action.

Transition from Individual to Collective Learning

Building resilience to recurring risk at the community level requires shifting from individual to collective learning. Collective learning is defined as the simultaneous recognition of changing conditions that represent risk by large numbers of people, enabling them to take coordinated action to reduce that risk (Castells 2009). Collective learning is constrained by the same limitations as individual learning, but on a wider scale. Because larger numbers of people are involved but time is urgent, the process necessarily requires a technical information infrastructure to support the search and exchange of information that enables learning on a broad scale. Designing the form and timing of communications through a technical information infrastructure creates the possibility of shaping the collective learning process.

Concepts from research in neuroscience that portray how the human brain learns new skills offer insights that are relevant to the larger question of how communities learn in a collective social setting (Edelman 1987; Arena and Michelon 2018). Researchers recognize that experiences change neural networks in the brain but also that experiences may lead to positive or negative changes in behavior. Understanding the process by which the content of experience is communicated among neural networks in the individual brain, leading to behavioral change, and distinguishing constructive from destructive learning patterns in practice are continuing research challenges. We propose that the concept of neuroplasticity that characterizes the process of learning new behaviors observed in an individual brain extends to the collective learning processes observed in social networks, if information is considered the energy that activates learning (Smith 2008; Costandi 2016; Comfort 2019).

The challenge is to identify key actors who initiate social change and devise means of communication to engage those actors in social networks to enable collective learning in a community exposed to recurring risk. In a social context, communication occurs among humans aggregated in social organizations or groups that generate processes of information search and exchange throughout a community at risk from recurring hazards. Mapping the relationships among groups that engage in social learning and the processes that characterize information flow within a given community provides the basis for understanding its capacity for collective learning and action and for redesigning the information flow to achieve the wider goal of risk reduction more effectively.

Enabling Collective Action for Communities at Risk

The centrality of communications to effective mitigation and response in natural disasters has long been recognized by practicing emergency managers and experienced hazards researchers (Mileti 1999; Waugh 2000; Comfort 2007), yet the link between communication and action is never certain. Recognizing what factors enable collective action—and, conversely, what conditions impede the process—is critical for building resilience in communities exposed to recurring risk (Comfort 2007). Three factors interact to create the possibility for enhancing communications on a community-wide scale: initial awareness of risk in the community, time available to avert harm, and existing technical infrastructure to support communications. The interaction among these three factors shapes the process of communication in actual disaster operations that either engenders collaboration to bring the incident under control or falters in dysfunction, triggering cascading disruption and damage (Comfort and Haase 2006). Communications serves a fundamental role in determining the level of interorganizational performance in a community exposed to recurring risk. This process warrants careful review, redesign, and investment to achieve the goal of community resilience to extreme events.

Four questions address the performance and structure of communications in crisis environments:

1. What are the primary characteristics of an interactive communications process between government personnel and community residents to reduce disaster risk?
2. What means of communication increase the capacity of community residents to recognize risk and act collectively to reduce harm?
3. What factors inhibit the capacity of community residents to recognize risk and act collectively to reduce harm?
4. What benefits or costs accrue to the community from implementing an information technology infrastructure to enable interactive communication in managing recurring risk?

We explore these four questions and propose a strategy to increase timely, informed collective action in risk-prone environments through use of a novel, technology-supported communications network at the neighborhood level to reduce disaster risk.

Communications in Disaster Environments

Increasing the level of knowledge and relevant experience among both emergency personnel and community residents becomes a major investment in building resilience to recurring risk. Sudden-onset events, such as earthquakes, tornadoes, wildfires, and tsunamis, create situations of urgent stress and surprise that challenge both practicing managers and community residents in mobilizing effective action in response to an actual event. In environments of known risk, investments in communications infrastructure, training, and practice can reduce the surprise element in hazardous events and increase the capacity of community residents to recognize sudden risk more quickly and act more effectively on incoming information regarding an emerging threat. In practice, preparedness and training exercises for hazardous events reduce the cognitive load for both emergency personnel and community residents, enabling them to exchange and process valid information regarding hazards more easily and focus limited attention on the most urgent needs (Mayer and Pilegard 2021, 241).

To increase the effectiveness of communications in disaster environments, we have developed an information technology platform that uses multiple means of representing information—maps, diagrams, written words, and visual images—in a process of information search, exchange, and iterative learning. This platform, called the community resilience framework (CRF), provides a common operational profile of the community, integrating data from multiple sources to create an overall assessment of the status of risk to the community at any given time.

The design of the CRF draws on four conceptual sources that bear directly on the problem of mobilizing collective action under threat. Each set of concepts focuses on a critical stage of this process: perception, recognition, transmission, and communication of risk. If the process fails at any one stage, the capacity for mobilizing collective action is limited. First, the limited "absorptive capacity" of human decision makers decreases in complex environments, constraining perception (Cohen and Levinthal 1990). This condition is especially critical in urgent conditions as the rapid flow of information exceeds the human capacity to absorb and understand new information and, at a certain threshold level of exposure, human cognitive processes simply shut out information that is too complex or different from previous experience. Consequently, human managers or community residents may receive information about risk but fail to perceive its implications for action. This limitation is particularly acute in the context of extreme events. This first stage, denial of risk, was readily apparent among many actors in the events leading up to the 2004 Sumatra earthquake and tsunami.

The second stream of inquiry addresses the capacity of individuals to recognize risk under urgent conditions (Weick 1993, 1995; Weick and Sutcliffe 2001, 2015). To Weick, this capacity is a process of gleaning cues from a changing environment and interpreting them as a basis for action. To generate collective action, individuals need to recognize the cues for danger, create meaning from the cues to enable action, and communicate that meaning to others. The inability to achieve such action under threat lies most frequently in the individual's inability to translate the meaning of emerging danger into timely communication that is understood by other members of the group or community. When that occurs, the link between individual perception of risk and group comprehension is broken. The inability of Max Mayfield, director of the US National Weather Service, to translate the danger of the 2005 Hurricane Katrina strengthening in the Gulf of Mexico to local and state officials in Louisiana in terms they could understand illustrates this breakdown between communication and action on several administrative levels (ABC News 2005; Strohm 2005).

Third, timely and valid transmission of risk to others in immediate danger builds the capacity for collective action. Klein and colleagues (1993) acknowledged that, under threat, the process of reasoning through a linear set of instructions or rules was far too slow to avoid danger. Rather, they observed that experienced leaders drew on previous situations of similar conditions and created workable strategies based on their previous experience to fit the existing conditions more appropriately. The limits of this approach lie in the maxim, paraphrasing Simon (1962, 479) that, as humans, "we can only create what we already understand." If actors under threat confront a situation so completely different from their previous experience, they find little meaning that could serve as a basis for action. The set of events confronting Japanese decision makers in the 2011 Tohoku disasters illustrates this paralysis in action.

Leadership in decision making, a fourth set of concepts, shifts from a focus on individual performance under stress to building a team that can function coherently in extreme danger. Many organizations have adopted this strategy to cope with unexpected, challenging events, such as the Federal Emergency Management Agency's strategy of "Unified Command" incorporated into the National Incident Management System (FEMA 2017). Although military units, firefighting crews, and other organizations working in risk-prone contexts have long relied on team building and training to hone the capacity for coordinated action in extreme danger (Flin et al. 2008; Nowell et al. 2018), this type of training is often designated for specific types of risk. A decision model aptly termed "highly optimized tolerance" (HOT) developed by Carlson and Doyle (1999, 2000, 2002) combines design and deep knowledge of specific environments with spontaneous action to achieve robust system performance. Designed initially to provide decision support to firefighting teams in rapidly moving wildlands fires in California, HOT has proven effective for the specific type of risk for which it is designed, but it becomes vulnerable to error in unexpected, unknown conditions.

These four sets of concepts underscore the key premise that adaptation in disaster environments represents a form of learning, first by individuals and second as individuals transmit insights gained to others and foster a process of collective learning among groups. In a third iteration, groups then extend the learning process to other groups that, by design, includes all members of the community. If communication is understood as a process of iterative learning, the modes and technical means of communication can be designed to facilitate learning for the community.

In its classic form, communication constitutes a two-way or multiway process between sender(s) and receiver(s) (Shannon 1948; Wiener 1948). Effective communication relies on the shared understanding of terms and concepts exchanged between sender and receiver that characterize the operating environment. Increasing familiarity with the working vocabulary of terms, concepts, and visual references used to characterize the operating environment allows both senders and receivers to focus their cognitive processing on evaluating and assimilating new messages, rather than using limited working memory to understand unfamiliar terms (Mayer and Pilegard 2021). Prior related knowledge confers an ability to recognize the value of new information, assimilate it, and apply it to practical ends. These abilities, taken together, constitute "absorptive capacity" for individuals and, collectively, for a community (Cohen and Levinthal 1990).

Designing an Information Infrastructure to Support Collective Learning

Information technology offers a promising means of extending human capacity to recognize, comprehend, and act in situations of extreme danger and to create a shared basis for collective action. Information technology can facilitate the tasks outlined briefly above. Computational design of "views" of information specific to each manager's responsibility in a complex environment reduces the problem of overload and enables each manager to comprehend the impact of an event in a particular area or context. In synthesizing disparate cues from a changing environment, a "bowtie" architecture elicits information from multiple sources and analyzes it from the perspective of the whole community, strengthening the capacity of decision makers to integrate different types of information in a coherent profile of risk; see figure 1-1. The bowtie architecture transmits relevant information to responsible agencies in a simultaneous process that enables each agency to take informed action in a self-organizing but coherent manner. Recurring sensing from the network picks up changes in the environment, analyzes the aggregate information from

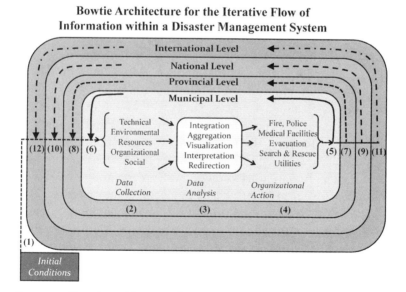

FIGURE 1-1. Bowtie architecture for multilevel information flow in crisis management.

the perspective of the whole community, and transmits synthesized information to relevant decision makers as updates for iterative analysis at scalable levels of action (Comfort et al. 2009). Representing such updates in well-designed visual graphics enhances human capacity to absorb difficult information in stressful contexts.

Factors that impede broad, community-wide recognition of risk in extreme events include heterogeneity among participating organizations and groups, asymmetry in information processes among the groups, and asynchronous dissemination of critical information to participating groups. For each of these factors, which are difficult for human managers or organizations to overcome, information technologies offer technical means for providing decision support. Providing focused, timely, and valid information to diverse sets of actors is critical to mobilizing collective action for the community at risk.

Bowtie architecture for information infrastructure

Creating an aligned understanding about response operations among decision makers and responders activates the exchange of information that enables actors at different levels of responsibility to take coherent action simultaneously (Comfort 2007). In environments of intense and rapid information flow, technical support can facilitate this alignment of relevant information among different jurisdictions and funding sectors. The process is necessarily iterative and involves three sequential steps in managing information flow: selection, organization, and integration (Axelrod and Cohen 1999; Azevedo 2014). These core functions contribute to reducing cognitive load for decision makers and, when designed skillfully for specific contexts, can be implemented through computational means.

Figure 1-1 illustrates the bowtie architecture for information flow across multiple levels of jurisdictions in an iterative process. This design channels information from different sources through an analytical "knot" or center that sorts and organizes incoming information, analyzes it in terms of existing knowledge and goals for the community, and disseminates relevant information to specific agencies or groups to support timely and coordinated decision making among multiple actors.

Communication flow across multiple levels

Exchanging core information among major responders during a crisis determines the capacity for collective action over the course of disaster response operations (Corbacioglu and Kapucu 2006). Core information includes both organizational structure for decision making and constraints that depend on context. An appropriate information infrastructure supports essential search and exchange functions among participating actors in crisis operations and facilitates the flow of core information among them (Comfort et al. 2004). By simulating the flow of information during a tsunami event at multiple levels of action—government decision makers, community leaders, and community followers—we test the effectiveness of a prototype sociotechnical communications infrastructure designed to increase collective action among the participants. The next step is to design a computational model that uses a range of means of communication to facilitate the flow of information through a network of multiple organizations and actors at the community level, using a sociotechnical information infrastructure.

Community Resilience Framework

To enable collective adaptation in a changing context, we designed a software platform, the community resilience framework (CRF), to support collaborative decision making among organizations. The CRF platform is a technical implementation of the iterative information flow outlined in the bowtie architecture above. It integrates widely used information technology (IT) services and devices, using professional decision-making protocols and individual smartphones, in a multimedia communication design. This design allows users to communicate through different types of IT devices to overcome both cognitive limitations and physical distance. A detailed account of the CRF, its design, components and functions, is presented in chapter 3.

The CRF is designed for the disaster management environment using principles drawn from cognitive and educational psychology research as well as disaster management protocols. Much of the research in cognitive psychology has examined the relationship between learner performance and organizations, measured by density of learners, type of organization, spatial orientation of learners and organizations, and temporal rendering of essential information. Within each level of the CRF, specific principles of cognitive load management have been implemented to select and organize information so that it may be integrated more accurately and rapidly into the operating environment of individual, group, and community users. As illustrated in figure 1-1, the bowtie architecture receives information from multiple sources for the whole community at risk but depends on focusing essential information at each jurisdictional level. This approach uses fast learning, or system 1 (Kahneman 2012), to detect novel changes in incoming data but checks that information against stored knowledge from the community using slow learning or system 2 (Kahneman 2012) to constrain incoming information to a context-essential flow. This approach reduces extraneous cognitive load for senders and receivers at individual, group, and collective levels, enabling actors at each level to comprehend the larger threat to the whole community more quickly.

The combination of cognitive load reduction and designed data views develops the rapid flow of information in each of the CRF modules described in subsequent chapters 3, 4, and 7 that connect the undersea and community-based communications infrastructure with the social and leader networks. These network-centric modules focus cognitive resources on specific information flows that have been reduced to only the essential, intrinsic loads required by the role and jurisdiction of the actors. A data view may comprise one or many roles or jurisdictions depending on the actor, organization, and context. Views may be constructed from one or many other perspectives, allowing reusable, logical objects to be deployed across jurisdictions and organizational roles. These views rapidly provide necessary information to community groups, allowing self-organization while reducing errors prevalent in high-stress, high-data-volume disaster environments. The redesigned information flow uses consistent and familiar symbols, controls, and visualizations to ensure a common operating profile for actors in all roles and jurisdictions. The technical implementation of ad hoc, opportunistic networks is discussed in detail in chapters 3 and 4.

Selection, organization, and integration: The SOI model of information processing

Individual responders must act simultaneously through real-time exchange of information to adapt organizational performance collectively under changing conditions. Each communication device included in the CRF contributes to the selection, organization, and integration (SOI) model of combining information from multiple sources to interpret rapidly changing conditions (Azevedo 2014). *Selection* helps people focus on relevant pieces of incoming information from multiple sources, *organization* helps people order the incoming information into a coherent representation for a specific context, and *integration* allows people to activate and use prior knowledge to interpret incoming information and to integrate multiple views of conditions at risk. These three components of the SOI model are central to the design necessary for reducing extraneous load for individual cognitive processes in multimedia formats.

The intents of the SOI model are to reduce cognitive time needed to arrange disparate material for rapid absorption under stressful conditions and to maximize cognitive time available for innovative design of strategies for action in urgent disaster environments. Much of the cognitive processing required for building a successful learning outcome occurs in working memory (system 1) (Mayer and Fiorella 2014). If actors are confronted with unfamiliar or startling events, they may not be able to engage in the rapid cognitive processing needed to make sense of the critical information (Mayer and Fiorella 2014).

Increasing collective problem-solving capacity

Cognitive capacity is reduced when working memory is overwhelmed with processing extraneous information. Mayer and Fiorella (2014) discuss principles of a cognitive theory of multimedia learning (CTML) that reduce the amount of extraneous information processing that occurs in multimedia environments. These principles include coherence, signaling, redundancy, spatial contiguity, and temporal contiguity. When incorporated into the design of systems using data from different sources presented in varied formats, these principles increase the cognitive capacity of participants by reducing the duration of time required for processing critical information in limited working memory. By structuring and visualizing information according to the CTML, participants identify critical information more rapidly and process it more accurately, allowing additional data to be received into working memory. Three principles most relevant to the design of the CRF are the coherence principle, the signaling principle, and the spatial contiguity principle.

The coherence principle. "People learn more deeply from a multimedia message when cues are added that highlight the organization of essential material" (Mayer and Fiorella 2021, 145). In figure 1-2, the coherence principle is illustrated using the CRF Twitter Map visualization. The visualization presents a specific region in the Padang City area to the viewer. The screen is divided into three areas: the left-hand menu, the upper menu, and the main content area. The left-hand menu contains only essential navigational elements necessary for operating the CRF. Similarly, the upper menu contains minimal navigational elements. The

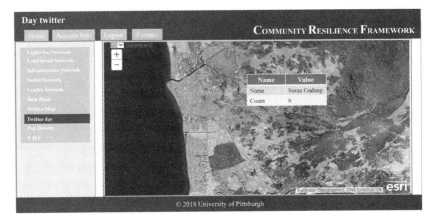

FIGURE 1-2. Coherence principle: Illustration of visualization on CRF Twitter map.

main content area is the focal point on which the viewer should be focused to receive information about the population in the Surau Gadang sub-subdistrict. No additional information outside of the critical informational elements needed to establish coherence regarding the Twitter population is presented to the viewer. This visualization allows for rapid selection, organization, and integration of information by the viewer, reducing the need to focus limited working memory on extraneous information and freeing cognitive processes (system 1 and system 2) to engage with new information.

The signaling principle. "People learn more deeply from a multimedia message when cues are added that highlight the organization of the essential material" (Mayer and Fiorella 2021, 221).

In figure 1-3, the evacuation map visualized on the leaders' handheld devices displays all the routes available to the leaders in the field guiding followers to shelters. This visualization does not use the signaling principle, requiring the leader to expend additional cognitive processing effort while attempting to select relevant information and determine which routes are most relevant and appropriate to follow in guiding followers to safety. Similarly, in figure 1-4, decision makers at the emergency operations center (EOC) will experience an increased cognitive load as they encounter the area map in the CRF, with which they may lack familiarity. This novel information will likely generate an increased cognitive load to process critical information for timely, efficient decision making.

In figure 1-5, the evacuation map again presents all the routes available

FIGURE 1-3. Available routes for evacuation, without signaling, leaders' view.

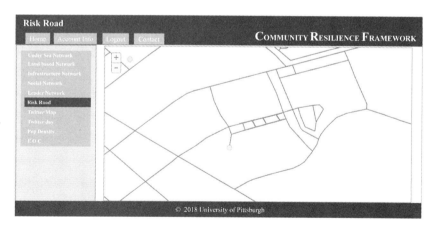

FIGURE 1-4. Recommended evacuation route, without signaling, EOC personnel view.

to the leader. Now, however, the leader is directed toward the relevant evacuation map segments using darker gray coloration as a visual signal for their follower group. The amount of cognitive processing necessary to select, organize, and integrate relevant information is sharply reduced. Similarly, EOC personnel will also experience a decreased cognitive load as they will be able to process essential material faster. This change will potentially have a larger effect in reduction of cognitive processing for the EOC decision makers because many may be less familiar than the community leaders with the districts, subdistricts, and sub-subdistricts represented in the risk road screen shown in figure 1-5.

Figure 1-6 displays the evacuation path visualization, demonstrating a similar signaling mechanism used to focus viewers on the essential information being displayed, allowing them to select and integrate relevant portions of messages as they assess evacuation paths. The evacuation path visualization represents basic road information, without additional,

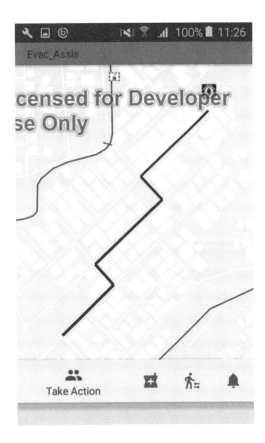

FIGURE 1-5. Evacuation map, with signaling, leaders' view.

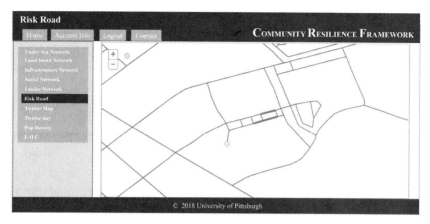

FIGURE 1-6. Evacuation map, with signaling, EOC personnel view.

possibly distracting, points of interest, shapes, or colors. Only the evacuation path segment relevant to the particular leader–follower group is rendered in darker gray, allowing immediate focus to be placed on this essential information and reducing the cognitive processing demand for the viewer.

The spatial contiguity principle. "People learn more deeply from a message including multiple types of information when corresponding words and pictures are presented near rather than far from each other on the page or screen" (Mayer and Fiorella 2021, 337).

A third principle, shown in figure 1-7, demonstrates the effect of spatial contiguity in the visualization of the undersea network. In this

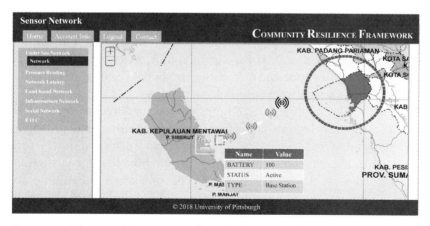

FIGURE 1-7. Close spatial representation of information for rapid learning.

undersea network visualization, only essential information is rendered using icons along the route of the cable. Similarly, a table of context-relevant text data is displayed in the same visual field as the map as a way to reduce extraneous processing by guiding the learner's cognitive processing (Mayer and Fiorella 2014). Although there is additional information available for visualization on the map, coherence is preserved by eliminating extraneous notations and elements, helping the viewers focus on critical undersea network information for the particular function in which they are engaged. In addition, the signaling and spatial contiguity principles are actively reducing other interacting sources of information that would potentially generate extraneous load and distract attention from the viewer's processing of essential information.

Testing a Conceptual Design in an Actual Risk Environment

This chapter has defined and characterized the problem of collective cognition of risk and the transition to collective action to reduce imminent danger. It has also outlined a conceptual model of the dynamic flow of information among individuals, organizations, and jurisdictions through the community resilience framework that builds collective cognition and activates collective response to threat. The practical task of implementing the conceptual components of the community resilience framework based on the CASoS model is an iterative process involving a range of actors and jurisdictions set in the physical context of risk. To build a community resilience framework that can be implemented in an actual risk environment requires careful design, implementation, and testing under field conditions that allow learning and adaptation in both the technical and organizational components of the system. The context selected for this field study is Padang, Indonesia, a city of approximately one million residents located on the western shore of Sumatra, Indonesia, exposed to major risk from near-field tsunamis.

Based on the organizational design of bowtie architecture for a working information platform and incorporating cognitive load principles into that design, the conceptual model of an information infrastructure outlined in this chapter creates a learning environment for facilitating rapid information flow among different types of actors at the community level. This learning environment depends on both technical and social/ organizational components to achieve its intended goal of building resilience in a community exposed to known risk.

The chapters in this book represent different iterations of the CASoS framework in reference to the same large-scale and complex policy problem: early detection and warning of near-field tsunamis. Importantly, the chapters illustrate the CASoS framework in the design, modeling, and testing of networks in actual environments at risk. Chapter 2 presents the risk assessment undertaken by an actual community, Padang, West Sumatra, Indonesia, as it organizes a program of community-wide preparedness to mitigate the risk of near-field tsunamis. Chapter 3 translates the conceptual design for interactive communication at the neighborhood level into a working software program to support collective action in an actual field environment, the city of Padang. It provides an overview of the technical components that make up a working information infrastructure; these components include the community knowledge base and information platform, the community resilience framework (CRF), with its central coordinator and connecting handheld mobile devices.

Chapter 4 extends the technical design of the CRF to include interactive communication among community residents during the dynamic conditions of evacuation, using an innovative prototype wireless network that can function in a disaster-degraded environment and guide residents to safety in the short 20- to 30-minute interval before the tsunami wave strikes. It also includes a simulation of the number of devices that can be supported by each segment of the wireless communication chain as neighborhood residents move safely under a network activated for communication in disaster contexts.

Chapter 5 introduces the concept and design of community-built shelters using neighborhood mosques as the initial base for collaborative construction and proposes these community shelters as the safest destinations for neighborhood residents in evacuation. Chapter 6 probes more deeply the concept of collective cognition as it shapes organizational performance and reports observations from field trials of a community-based network of electronic devices to support interactive communication among local emergency services personnel and community residents exposed to tsunami risk. It links the social/organizational network of community residents to the technical infrastructure and simulates the flow of information through the community using the technical network designed to support organizational decision making in a disaster-degraded environment. Chapter 7 presents results from experimental field tests of the sociotechnical framework conducted in both Padang and Pittsburgh. The tests assessed the rate of change from cognition to action among community residents using the sociotechnical network for transmission of information in comparison to standard means of communication. Chapters 8 and 9 present the design, experiment, and early findings from the undersea network that detects a tsunami signal using a seafloor sensor. Chapter 10 reports the actual deployment of the undersea network for the early detection and warning of near-field tsunamis in the Mentawai Sea. Chapter 11 examines the potential contribution of the validated working prototype to the Indian Ocean Tsunami Warning and Mitigation System and its twenty-four member nations that rim the Indian Ocean. Chapter 12, the final chapter, looks forward to the possible extension of tsunami early detection and warning systems to other areas of Indonesia or other coastal cities at risk from coastal hazards. Cumulatively, the set of twelve chapters illustrates the power of the CASoS framework in identifying the nexus between learning and action in urgent environments for community residents.

Conclusion

This chapter lays out a bold vision, but a demanding task, to increase the capacity for collective action in communities exposed to near-field tsunami risk. This vision acknowledges the potential of current information technologies to support the activation of local community networks in near-real time decision making and action. To do so, however, is not trivial. Four basic steps are essential in this interdisciplinary and international effort.

First, integrating technologies into everyday use by community members requires organizational design, cultural acceptance, and acquisition of communication skills. This step involves building deep knowledge of the local community, its policies regarding risk, the social/demographic characteristics of the population, the residents' general level of understanding of the risk to which the community is exposed, and their willingness to learn new concepts and develop operational skills to use new technologies.

Second, articulating a bold vision requires mastery of the technical skills and electronic devices that are readily available to community residents and the design of appropriate software applications that enable community residents to understand their functions easily and develop the skills needed to use the devices in urgent conditions. Designed and tested successfully at the community scale, wireless networks designed for disaster-degraded environments will enable an advanced level of communication and rapid information search and exchange that strengthen known networks of social organization and cultural beliefs to protect and sustain continuity of operations.

Third, multiple methods of inquiry and analysis are essential. Technologies assist, but do not replace, collective learning. The focus necessarily remains on the human actors in the system and their capacity to learn new skills and integrate a broader range of risk information into everyday practice.

Finally, identifying the threshold point of change in the community action system is critical—that is, the point at which collective recognition of risk to the community shifts directly into collective action. In the language of complex systems, this step represents a phase transition that marks a dramatic shift in the performance of the whole system (Holland 1996; Axelrod and Cohen 1999; Solé 2011). This threshold of change will likely shift many times in interaction with other forces of economic, political, and cultural change in the society and will need to be recalibrated as external threats impinge on the internal capacity of organizations and jurisdictions to manage productive operations for the community over time. Importantly, it is a measure of how quickly information regarding risk flows through the community, leading to adaptive change in practice as individuals, groups, organizations, and jurisdictions learn to live with risk more effectively.

The sociotechnical designs, experiments, and simulations presented in chapters 3, 4, 6, and 7 provide detailed computational models and simulated tests of performance for the set of components that make up a community resilience framework. In practice, the framework will foster community resilience to sudden, urgent threats in complex, urgent environments.

References

ABC News. 2005. ABC News Network, December 15, 2005. https://abcnews.go.com/WNT/PersonOfWeek/story?id=1153294.

Arena, Claudia, and Giovanna Michelon. 2018. "A matter of control or identity? Family firms' environmental reporting decisions along the corporate life cycle." *Business Strategy and the Environment* 27, no. 8: 1596–1608.

Argyris, Chris, and D. Schön. 1996. *Organizational Learning II: Theory, Method, and Practice.* Addison-Wesley.

Axelrod, Robert M., and Michael D. Cohen. 1999. *Harnessing Complexity: Organizational Implications of a Scientific Frontier.* New York: Free Press.

Azevedo, Ana, ed. 2014. *Integration of Data Mining in Business Intelligence Systems.* IGI Global.

Beck, Ulrich. 1992. *Risk Society: Towards a New Modernity.* Translated by Mark Ritter. London: SAGE.

Carlson, Jean M., and John Doyle. 1999. "Highly optimized tolerance: A mechanism for power laws in designed systems." *Physical Review E, Statistical Physics, Plasmas, Fluids, and Related Interdisciplinary Topics* 60: 1412–27.

Carlson, Jean M., and John Doyle. 2000. Highly Optimized Tolerance: Robustness and Design in Complex Systems. *Physical Review Letters* 84: 2529–32.

Carlson, Jean M., and John Doyle. 2002. "Complexity and robustness." *Proceedings of the National Academy of Sciences* 99: 2538–45.

Castells, Manuel. 2009. *Communication Power*. Oxford University Press.

Cohen, Wesley M., and Daniel A. Levinthal. 1990. "Absorptive capacity: A new perspective on learning and innovation." *Administrative Science Quarterly*. 128–52. Retrieved from https://search .proquest.com/docview/203983133?accountid=10610.

Comfort, Louise K. 2007. "Crisis management in hindsight: Cognition, communication, coordination, and control." *Public Administration Review* 67: 189–97.

Comfort, Louise K. 2019. *The Dynamics of Risk: Changing Technologies and Collective Action in Seismic Events*. Princeton University Press.

Comfort, Louise K., and Thomas W. Haase. 2006. "Communication, coherence, and collective action: The impact of Hurricane Katrina on communications infrastructure." *Public Works Management and Policy* 10, no. 4: 328–43.

Comfort, Louise K., Kilkon Ko, and Adam Zagorecki. 2004. "Coordination in rapidly evolving disaster response systems: The role of information." *American Behavioral Scientist* 48, no. 3: 295–313.

Comfort, Louise K., Daniel Mosse, and Taieb Znati. 2009. "Managing risk in real time: Integrating information technology into disaster risk reduction and response." *Commonwealth: A Journal of Political Science* 15: 27–46.

Comfort, Louise K., Daniel Mosse, Taieb Znati, and Thomas W. Haase. 2007. "Socio-Technical Aspects of Cost-Effective, Sensor-Based Infrastructure for Tsunami Detection and Monitoring." Research Report. Pittsburgh, PA: Center for Disaster Management, University of Pittsburgh.

Corbacioglu, Sitki, and Naim Kapucu. 2006. "Organisational Learning and Self Adaptation in Dynamic Disaster Environments." *Disasters* 30, no. 2: 212–33.

Costandi, Moheb. 2016. *Neuroplasticity*. MIT Press.

Edelman, Gerald M. 1987. *Neural Darwinism: The Theory of Neuronal Group Selection*. New York: Basic Books.

FEMA (Federal Emergency Management Agency). 2017. *National Incident Management System*, 3rd ed. Washington, DC: US Department of Homeland Security.

Glass, Robert J., Arlo L. Ames, Theresa J. Brown, Walter E. Beyeler, Patrick D. Finley, John M. Linebarger, Nancy S. Brodsky, et al. 2011. *Complex Adaptive Systems of Systems (CASoS) Engineering: Mapping Aspirations to Problem Solutions*. No. SAND2011-3354C. Albuquerque, NM: Sandia National Laboratories.

Flin, Rhona, Paul O'Connor, and Margaret Crichton. 2008. *Safety at the Sharp End: A Guide to Non-Technical Skills*. Boca Raton, FL: CRC Press/Taylor and Francis Group.

Holland, John H. 1996. *Hidden Order: How Adaptation Builds Complexity*. New York: Basic Books.

Kahneman, Daniel. 2012. *Thinking, Fast and Slow*. New York: Farrar, Straus and Giroux.

Klein, Gary A., Judith Orasanu, Roberta Calderwood, and Caroline E. Zsambok. 1993. *Decision Making in Action: Models and Methods*. Norwood, NJ: Ablex.

Landau, Martin. 1969. "Redundancy, rationality, and the problem of duplication and overlap." *Public Administration Review* 29, no. 4: 346–58.

La Porte, Todd R. 1975. *Organized Social Complexity: Challenge to Politics and Policy*. Princeton, NJ: Princeton University Press.

Luhmann, Niklas. 1996. *Ecological Communication*. University of Chicago Press.

Martin, Lisa, and Naaman Zhou. 2018. "Indonesia Tsunami Caused by Collapse of Volcano." *The Guardian*. December 24, 2018. https://www.theguardian.com/world/2018/dec/24/sunda -strait-tsunami-volcano-indonesia.

Mayer, Richard E., and Celeste Pilegard. 2021. "Principles for managing essential processing in multimedia learning: Segmenting, pretraining, and modality principles." *Cambridge Handbook of Multimedia Learning*, 3rd ed. Cambridge, UK: Cambridge University Press, 241–74.

Mayer, Richard E., and Logan Fiorella. 2021. *Cambridge Handbook of Multimedia Learning*, 3rd ed. Cambridge, UK: Cambridge University Press.

Mileti, Dennis. 1999. *Disasters by Design: A Reassessment of Natural Hazards in the United States*. Washington, DC: Joseph Henry Press.

Murphy, Gillian, John A. Groeger, and Ciara M. Greene. 2016. "Twenty years of load theory—where are we now, and where should we go next?" *Psychonomic Bulletin and Review* 23, no. 5: 1316–40.

Nowell, Branda, Toddi Steelman, Anne-Lise K. Velez, and Zheng Yang. 2018. "The structure of effective governance of disaster response networks: Insights from the field." *American Review of Public Administration* 48, no. 7: 699–715.

Paas, Fred, and John Sweller. 2014. "Implications of cognitive load theory for multimedia learning." In *Cambridge Handbook of Multimedia Learning*, 2nd ed., ed. Richard E. Mayer and Logan Fiorella. New York: Cambridge University Press, 27–42.

Rittel, Horst W. J., and Melvin M. Webber. 1973. "Dilemmas in a general theory of planning." *Policy Sciences* 4, no. 2: 155–69.

Roberts, Karlene H. 1993. *New Challenges to Understanding Organizations*. Macmillan.

Shannon, Claude E. 1948. "A mathematical theory of communication." *Bell System Technical Journal* 27, no. 3: 379–23.

Simon, Herbert. 1962. "The architecture of complexity." *Proceedings of the American Philosophical Society* 106, no. 6: 467–82.

Simon, Herbert A. 1996. *The Sciences of the Artificial*. Cambridge, MA: MIT Press.

Smith, Eric. 2008. "Thermodynamics of natural selection I: Energy flow and the limits on organization." *Journal of Theoretical Biology* 252, no. 2: 185–97.

Solé, R. V. 2011. *Phase Transitions*. Princeton, NJ: Princeton University Press.

Strohm, Chris. 2005. "Weather service officials gave dire, accurate warnings before Katrina hit." *Government Executive*. September 22, 2005. https://www.govexec.com/defense/2005/09/weather-service-officials-gave-dire-accurate-warnings-before-katrina-hit/20228/.

Sweller, John. 1988. "Cognitive load during problem solving: Effects on learning." *Cognitive Science* 12, no. 2: 257–85.

USGS (United States Geological Survey). 2018. "Magnitude 7.5 earthquake near Palu, Indonesia, September 28, 2018." USGS Office of Communications and Publishing, September 28, 2018. https://www.usgs.gov/news/magnitude-75-earthquake-near-palu-indonesia.

Van de Walle, Steven. 2014. "Building resilience in public organizations: The role of waste and bricolage." *Innovation Journal* 19, no. 2: 1–18.

Waugh, William L., Jr. 2000. *Living with Hazards, Dealing with Disasters: An Introduction to Emergency Management*. New York: Routledge.

Weick, Karl E. 1993. "The collapse of sensemaking in organizations: The Mann Gulch disaster." *Administrative Science Quarterly* 38, no. 4: 628–52.

Weick, Karl E. 1995. *Sensemaking in Organizations*. SAGE.

Weick, Karl E., and Kathleen M. Sutcliffe. 2001. *Managing the Unexpected: Assuring High Performance in an Age of Complexity*. Jossey-Bass.

Weick, Karl E., and Kathleen M. Sutcliffe. 2007. *Managing the Unexpected: Resilient Performance in an Age of Uncertainty*, 2nd ed. Jossey-Bass, xii.

Weick, Karl E., and Kathleen M. Sutcliffe. 2015. *Managing the Unexpected: Sustained Performance in a Complex World*, 3rd ed. New York: Wiley.

Wiener, Norbert. 1948. *Cybernetics or Control and Communication in the Animal and the Machine*, 2nd ed. New York: Wiley. First published 1961, MIT Press.

Wright, Stephen. 2018. "Warning system might have saved lives in Indonesian tsunami." Phys.org, September 30, 2018. https://phys.org/news/2018-09-indonesian-tsunami.html.

Chapter 2

Community Networks for Tsunami Early Warning

Harkunti P. Rahayu and Louise K. Comfort

Near-field tsunamis present a recurring risk to coastal communities throughout the world but represent a massive risk to Indonesia and other countries that rim the Indian Ocean Basin. The deadly Indian Ocean tsunami of December 26, 2004, left nearly 230,000 dead, approximately 125,000 from Indonesia alone, and damage to infrastructure, housing, and livelihoods in cities and communities across the region exceeded more than US$7 billion.

At that time, the landscape of tsunami preparedness in Indonesia as well as in the Indian Ocean region was considered low. Although the region was characterized by tsunami risk, the danger was not well known. No tsunami early warning system existed; communities were unaware and not prepared for tsunamis. There was extremely limited response planning for tsunamis and lack of recovery planning for such mega-disasters.

Reducing Risk of Near-Field Tsunamis

After this devastating event, the Indonesian government developed the Indonesian Tsunami Early Warning System (InaTEWS) in an effort to alert people at risk, that is, coastal communities in tsunami-prone

This paper is an adaptation of an earlier paper by H. Rahayu, L. Comfort, R. Haigh, D. Amaratunga, and D. Khoirunnisa, 2020, "A study of people-centered early warning system in the face of near-field tsunami risk for Indonesian coastal cities," *International Journal of Disaster Resilience in the Built Environment* (January), doi:10.1108/ IJDRBE-10-2019-0068.

areas. Indonesia also joined regional efforts among countries that border the Indian Ocean to develop the Indian Ocean Tsunami Warning and Mitigation System (ICG/IOTWMS), which earned the support of the Intergovernmental Oceanographic Commission of United Nations Education, Scientific, and Cultural Organization (IOC-UNESCO) in 2005. Indonesia's program, InaTEWS, became operational and was officially launched in 2008 by the president of Republic of Indonesia.

Tsunami risk reduction represents an extraordinarily difficult geophysical, technical, social, and organizational problem, even with carefully developed plans. Then, in 2018, two tsunamis struck coastal cities in Indonesia, again creating havoc, death, and destruction. On September 29, 2018, the city of Palu on Sulawesi Island was struck by a tsunami caused by a coastal landslide induced by an earthquake (Kumar et al. 2019; PuSGeN 2019; Gunawan et al. 2020; Natawidjaja et al. 2021). This unusual event also generated interacting hazards—earthquake, tsunami, and liquefaction—taking more than 4,300 lives. Barely three months later, on December 22, 2018, a second tsunami occurred in the Sunda Strait due to collapse of a flank of the undersea volcano, Gunung Anak Krakatau (Solihuddin et al. 2019). The tsunami-stricken communities on the shores of both Sumatra and Java islands lost 426 lives and incurred injuries for more than 14,000 residents. Some people survived by following the direction of tsunami signage for evacuation, however (Rahayu 2019).

In both instances, InaTEWS was not able to alert the populations at risk in time to evacuate because the system was designed to predict potential tsunamis generated only by seismic movement caused by tectonic earthquakes, not by other sources. As a result, during the Twelfth Session of the Intergovernmental Coordination Group for the Indian Ocean Tsunami Warning and Mitigation System (ICG/IOTWMS-XII) meeting at Kish Island, Iran, in March 2019, Indonesia, together with India and Australia as the tsunami service providers, committed to increase the capacity of the tsunami warning system by focusing also on nonseismic-generated tsunamis, known as atypical tsunamis, as part of the system development. Scientific investigations after both 2018 tsunami events indicated that multiple factors contributed to the substantial losses in lives and property, including the high degree of tsunami risk, density of population in the coastal communities, lack of tsunami preparedness, and limitations of the existing tsunami early warning system.

Such deadly events compel the review and redesign of early tsunami warning systems. Little attention is paid when the early tsunami system warns people in time, but the dramatic reporting of deaths after the system fails to warn people at risk demands action. In such instances, the failure is more often attributed to missing institutional capacity rather than technological error (Garcia and Fearnley 2012). In complex warning systems, warnings detected at the source by technological instruments may not reach the community at risk if links are missing at the local level (Spahn et al. 2014) or if the content of the warning may be distorted through different dissemination means. Providing a good end-to-end communication system—from ocean source to people at risk—is vitally important.

The Indonesia Tsunami Early Warning System— A State of the Art

InaTEWS is an integration of natural, social, technical, and physical phenomena that aims to save as many lives as possible by alerting the *people at risk* with sufficient lead time to make decisions for evacuation (Rahayu and Nasu 2011). The system is complex and interactive, involving sophisticated technologies and multiple infrastructures, jurisdictions, and organizations, as well as networks of governance including public, private, and nonprofit agencies. The central tasks are to detect the tsunami hazard accurately and to transmit the flow of valid tsunami information quickly through the system so that a community at risk, guided by public actors, receives valid tsunami warning information and public alerts and can take timely action to move to a safe place. InaTEWS's data structure and information flow are composed of a structure (upstream) component and a culture (downstream) component. These two components involve multilevel government actors, as well as a wide range of regional, national, and local stakeholders (BMKG 2012); see figure 2-1.

The upstream component includes both a seismic monitoring system to detect earthquakes and a decision support system to disseminate tsunami early warnings, with support from the sea level monitoring system for updating the status of tsunami warning. The design, implementation, and management of technologies for the structure component is led by BMKG (Indonesian Agency for Meteorology, Climatology, and Geophysics) and includes participation of other national agencies, including BPPT (National Agency for the Assessment and Application of Technology) through a buoy system and BIG (Geospatial Information Agency) through a tide gauge system.

The culture component consists of more complex networks of gover-

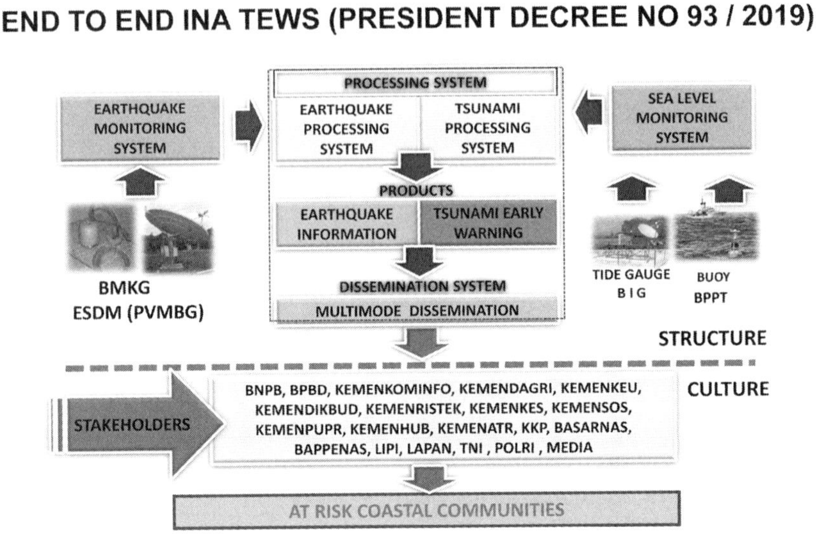

END TO END INA TEWS (PRESIDENT DECREE NO 93 / 2019)

FIGURE 2-1. Organizational diagram of InaTEWS, updated in 2019.

nance as shown in detail in figure 2-1. The downstream component constitutes a socio-technological system for transmitting tsunami warnings and includes the decision-making process of local governments for issuing evacuation orders or public alerts to the communities at risk, as mandated by the Indonesia Disaster Management Law (UU RI no. 24/2007), which defines the responsibility of local governments to protect the people. This public alert dissemination process is parallel with the tsunami warning dissemination using multimode communication infrastructures. Figure 2-2 shows the roles of the interface institutions, internet/GSM (Global System for Mobile Communication) providers and electronic media that provide significant support to disseminate the warning from BMKG directly to the communities at risk. The design, implementation, and management of the culture component depend on the policies, regulations, and standard operating procedures of tsunami warning, public safety management, technical infrastructure for information dissemination, and information sharing or exchange, as well as the degree of knowledge of tsunami risk and preparedness among coastal communities in responding to the warning.

Decentralization of the government system since 2000 in Indonesia has affected the diversity of local government priorities toward disaster risk-reduction countermeasures, including the warning system and

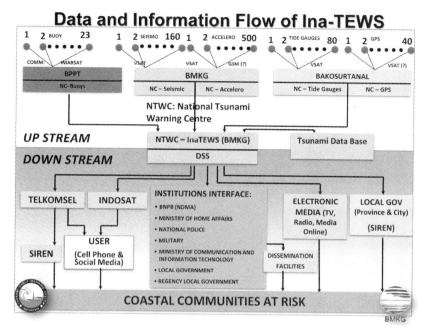

FIGURE 2-2. Data and information flow of InaTEWS (BMKG 2012).

preparedness, leading to wide variability in the culture component. Considering that 28 percent (146 out from 514) cities/regencies in Indonesia are exposed to high tsunami risk according to the National Action Plan for Disaster Risk Reduction (BNPB 2010), the downstream warning chain plays a very significant role in public alert and warning information dissemination to communities at risk.

In 2019, InaTEWS was legally strengthened by Presidential Decree no. 93/2019 to endorse the role of stakeholders for continuous improvement of the system; see figure 2-2. BMKG is still the lead agency for InaTEWS, with some improvement for the structure component—that is, earthquake monitoring. BMKG is supported by PVMBG (Center for Volcanology and Geological Disaster Mitigation–Ministry of Energy and Mineral Resources) for monitoring and detecting undersea volcanic activity. Sea level monitoring is supported by BPPT through a buoy system and by BIG through a tide gauge system. Since 2021, BPPT has been reorganized under BRIN (National Research and Innovation Agency).

The 2019 presidential decree strengthened more stakeholders' involvement in the culture component as interface agencies, covering almost all aspects of disaster risk reduction—*prevention, mitigation, preparedness,* and *response aspects*—to build resilient and responsive communities from all tsunami-prone regions. From the perspective of the stakeholders' role for *tsunami prevention and mitigation*, several institutions have significant responsibilities, they are BAPPENAS (Ministry of National Development Planning), KEMENATR/BPN (Ministry of Agrarian and Spatial Planning/National Land Agency), KEMENPUPR (Ministry of Public Works), KKP (Ministry of Marine Affairs and Fisheries), KEMENHUB (Ministry of Transportation), and KEMENDIKBUDRISTEK (Ministry of Education, Culture, Research, and Technology). In building *tsunami preparedness*, the roles of BNPB (National Disaster Management Agency), BPBD (Provincial/Local Disaster Management Office), LIPI (Indonesian Institute of Science), KEMENDAGRI (Ministry of Home Affairs), and media are very significant. For *tsunami evacuation and response*, the roles of KEMENKOMINFO (Ministry of Communication and Informatics), TNI (Indonesian National Armed Forces), POLRI (Indonesian National Police), and BASARNAS (National Search and Rescue Agency) are very important. The question remains regarding how to synergize effectively several national stakeholders stated in the decree that are not directly involved with local communities into the downstream component.

In 2021, however, there was a major reorganization of the governance structure, and this decree has not yet been updated. Like BPPT, two other agencies, LIPI and LAPAN, were also merged under BRIN, whereas KEMENRISTEK was merged as part of KEMENDIKBUDRISTEK (Ministry of Education, Culture, Research, and Technology).

Tsunami Risk in Indonesia and Prior Research and Development

Sumatra, Java, and Bali, the major islands of Indonesia, lie very close to the active subduction zone that runs parallel to the western coast of Sumatra and south from the Java, Bali, and Nusa Tenggara coasts. These coastal regions are highly exposed to the risk of near-field tsunamis—that is, a sudden ocean wave that is generated within 200 kilometers of shore by an undersea earthquake, landslide, or other seismic events. These events are especially dangerous given the brief time from detection to the arrival of a destructive wave on shore, typically twenty to thirty minutes for near-field events. Further, these coastal regions are densely populated, increasing the risk of harm to hundreds of thousands of residents. Given the history of destructive tsunamis in the region as recently as 2004 (Indian Ocean tsunami), 2007 (Pangandaran tsunami), 2010 (Mentawai tsunami), and 2018 (Palu and Sunda Strait tsunamis), Indonesia has focused significant attention, effort, and treasure to develop a viable tsunami early warning system.

After the devastating Indian Ocean tsunami of 2004, seventeen national agencies and institutions in Indonesia collaborated to develop InaTEWS, which was officially established in 2008 (BMKG 2012). Exploring models for early tsunami detection, the Indonesian design team studied the model proposed by the United Nations International Strategy for Disaster Reduction (UN-ISDR) in 2006 for a "people-centered early warning system." Such a system would include four components—risk knowledge, monitoring and warning service, dissemination and communication, and response capability (Basher 2006)—and they were adapted for InaTEWS. Rahayu (2012) has identified three of the four components as the most significant elements of the culture component: risk knowledge, dissemination and communication, and response capability. Risk knowledge and response capability focus on developing awareness, preparedness, and skills in the community, whereas dissemination and communication focus on the downstream warning chain (Rahayu 2012).

The monitoring and warning component, called the upstream element, is primarily technological and includes the detection systems, hardware, and decision support systems that identify the tsunami hazard. This component is implemented and managed by BMKG as the leading agency. The dissemination and communication component, or downstream element, communicates the tsunami hazard from BMKG through provincial and local governments to communities at risk using multimode communication infrastructure, such as a warning receiver system, fax, email, SMS (Short Message/Messaging Service), WhatsApp, Twitter, and television and radio broadcasts. This component is designed and managed by a solid, sound collaboration between BMKG and interface institutions, that is, BNPB, BPBD, KEMENKOMINFO, TNI and POLRI, KEMENDAGRI, and mainstream media such as television and radio. Local disaster management plays a very significant role not only in transmitting information for tsunami warning but also in sending the alerts or orders for evacuation to the communities at risk. The decision-making process for issuing the evacuation order is under the responsibility of local government as mandated by 2007 Disaster Management Law.

The responsibility for transmitting these two valid types of information (tsunami warning and order for evacuation) to community residents involves multiple institutions and actors that, working in collaboration, create a chain of timely, valid communication to warn of the tsunami hazard. This component requires building the capacity for collective action among multiple parties to enable timely evacuation in the very short time, twenty to thirty minutes, before a tsunami strikes shore. Given the dynamic change in communities, such systems need to be clearly established in policy and procedures, with education and training for community members, and regular exercises to maintain readiness for an unexpected tsunami event when it comes. Such policies will necessarily include evacuation away from the shore and to safe areas or shelters. Evacuation strategies need to be considered and planned, especially in dense urban areas, so that large numbers of people can move quickly to safe areas.

At the community level, response capability includes evacuation. Earlier studies of evacuation treated such behavior as individual activities. In time-urgent threats such as a tsunami in a densely populated area, collective action is critical as residents observe actions taken by others, receive information from external sources, and are strongly influenced by their prior knowledge and experience in disaster events, as well as their trust in the information received (Mawson 2005; Rahayu 2012).

The type and content of warnings regarding impending hazards significantly affect the degree to which they are heard, understood, and acted upon (Mileti and Peek 2000). Communication depends not only on the timing and content of the message, but also on the channel through which it is sent and the receptiveness of the receiver (Shannon 1948/1961; Wiener 1948; Comfort 2019). Creating the unity of sender, message, and receiver constitutes effective communication (Luhmann 1989/1996), and requires careful consideration of the characteristics of the community for which warning strategies are designed. The degree of risk awareness, trust in the source of the message—government agencies or media—and commitment to the community affect residents' acceptance of warning messages and their willingness to act on them. Given the range of possible influences on residents' receptiveness to hazard warnings and alerts, people-centered early warning systems need multiple methods and channels for disseminating warnings (Lindell and Perry 2012).

The question is, what methods are most effective, and how can they be developed? A study following the 2007 Padang earthquake found that 30 percent of the sample's respondents received tsunami warning information from informal sources (Taubenböck et al. 2009). Given the short time available in near-field events, this finding underscores the need to engage neighborhood leaders to facilitate the informal transmission of tsunami warning information to all community members. Engagement of community members in preparing for tsunami evacuations would likely increase local resilience and aid in recovery after an extreme event (Kapucu 2015).

Connecting the detection, or upstream, component with the dissemination, or downstream, component of an early warning system is critical (Sorensen 2000). To do so requires building this capacity through local neighborhood networks at the downstream level and integrating the informal community-centered approach with the hierarchical top-down management style of the upstream component (Thomalla et al. 2009). This leadership challenge crosses jurisdictional and organizational boundaries. Aligning formal power with informal power is a critical task in crisis management (Krackhardt 1990). It echoes the challenge of creating a "common operating picture" among all participants in response operations, which reflects collective cognition in practice (Comfort 2007). Creating an integrated, coherent early warning system that includes technical and human components involves developing a formal, written plan and conducting training exercises. These steps are critical when building a coordinated early warning system that includes a community-based network with emergent volunteers and acknowledges changing personnel in formal agencies over time (Whittaker et al. 2015).

The goal of a people-centered early warning system is to achieve end-to-end warning—from early detection of risk to household evacuation—to ensure that residents of neighborhoods, the "last-mile" population, receive timely, accurate warnings of imminent risk (Spahn et al. 2014). An effective, people-centered tsunami early warning system can only be realized if there are (1) *reliable* infrastructures to monitor, detect, analyze, and disseminate the warning of potential tsunami in accurate time; (2) *responsive and responsible* government officials at the local and regional level to receive tsunami warning and issue orders for evacuation in an orderly manner to the people at risk; and (3) *responsive* people (Rahayu 2014). Such a system requires sufficient technical capacity and clear operating procedures among multiple organizations, jurisdictions, and actors to achieve this goal. It values the connectedness of the actors but acknowledges the vulnerability of the sociotechnical information infrastructure through which communication flows. It is critical to envision the whole network in planning the downstream component of the early warning system.

Enhancing Disaster Preparedness and Root Causes of Societal Factors in Responding to Tsunami Warning

In line with the process development of InaTEWS from 2005 to 2008, strengthening the culture (downstream) component was done through the preparation of national tsunami exercises. There was a strong need to build a prepared and responsive community to accomplish the aims of the end-to-end tsunami exercise itself—that is, testing the readiness of the three strategic pillars of the tsunami warning system: (1) tsunami risk assessment and reduction; (2) tsunami detection, warning and dissemination; and (3) tsunami awareness, preparedness, and response (Rahayu et al. 2008)—and a standard operating procedure (SOP) for the end-to-end system was needed. Due to a wide range of variability at the local level, building SOPs for the downstream component was very challenging at that time because no such guidelines were used. Building public awareness and preparedness became necessary prior to the development of the SOP because process development of the SOP for the downstream tsunami warning chain and evacuation used a participatory, multistakeholders' approach.

In the case of Bali in 2006, the key step in building community preparedness was first working closely with community leaders, religious leaders, and both formal and informal leaders to build trust, followed by working with other community actors. These actors include households, school communities, community resilient groups (known as KSB), local nongovernmental organizations, and representatives of all strata. Preparedness was done through a community development program and included trainings, formal and informal meetings, and many other activities. Community development incorporated the participatory and capacity-building processes, using scientific judgment to identify the needs for external involvement, community focal points, and local knowledge and wisdom, as well as potential conflict areas from sociocultural, socioeconomic, political perspectives, and several others. The community development program used a three-tiered approach, starting by conducting a training for trainers exercise for about forty local actors as trainees. This training was followed by coaching the forty trainees to become facilitators directly in the communities of ten administrative villages called *Kelurahan* and traditional villages called *Banjar* located in the tsunami-prone area, with the target that every four trainees had to coach one hundred community residents for each village. The total number of people trained was one thousand from communities at risk. These trained community residents were expected to become "people getters" during tsunami drills and exercises. This three-tiered approach resulted in a total of about fifteen thousand participants who joined the Bali tsunami drill 2006, a successful achievement for building public awareness and preparedness within a short period (Rahayu et al. 2008). Activities during community development in Bali can be seen in the photo documentation shown in figure 2-3.

The community development activities for the tsunami drill in Cilegon, an industrial city in the primary industrial zone of Indonesia, in 2007 used a slightly different approach. The social demographics of the city document that almost 80 percent of the population were industrial workers in the high chemical, steel, processed timber, and other industries in which they have been trained in safety procedures to follow safety conduct strictly. The community development program and its SOP for evacuation were designed to anticipate multihazard events, that is, industrial hazards induced by earthquakes and tsunamis. The collaboration was not only with the Cilegon city stakeholders but also involved Bapeten (Nuclear Energy Regulatory Agency of Indonesia) and Nubika (Nuclear, Biological, and Chemical Agency of Indonesia, an institution under the army). The outcome of the tsunami drill was testing not only the three pillars of the tsunami warning system, but also the multihazard SOP for natural and technological disasters. To support many other cities and regencies of the tsunami-prone area, a guideline for tsunami drill preparation and implementation was developed based on lessons learned from the two tsunami exercises (Rahayu et al. 2008). The key issues in building community awareness and preparedness are the ability to know the level of risk and the ability to identify the right key actors of the penta-helix stakeholders: government, community, business sectors, academia, and media.

Since it was established in 2008, the tsunami warning system in Indonesia has been tested several times by real tsunami events, from minor to major. The most momentous event was the April 11, 2012, earthquake and tsunami event where natural signs of strong shaking and tsunami warnings created a chaotic situation; people panicked, and no proper evacuation process occurred either in Banda Aceh or Padang City, as if there were no tsunami preparedness intervention in place (Rahayu and

Figure 2-3. Community development activities. (Photos courtesy H. P. Rahayu.)

Nasu 2011). This event changed the approach and strategy of tsunami risk-reduction intervention by the development of the National Tsunami Disaster Risk Reduction Master Plan, 2012–2017.

Learning from these events and understanding the root causes of people's reactions, Rahayu and Kuraoka (2019) highlighted initial and important steps in improving the strategy of building and strengthening tsunami preparedness at the local, regional, and national levels to respond to tsunami warning:

- Lack of local tsunami awareness and preparedness indicates mainly that public education has not embedded the culture of public safety well into peoples' minds, as well as the short memory of people toward disaster events.
- Limited capacity of local government to disseminate the order for evacuation upon receiving a tsunami warning issued by national government (BMKG) is due to lack of standard operating procedures.
- Lack of responsive communities to tsunami warnings and orders for evacuation.
- Difficulties in planning and implementing tsunami disaster risk reduction due to the largely autonomous operation of government agencies at the local level and the diversity of ethnic cultures.

An in-depth study of the communities' response to tsunami warning in Padang City during the 2009 event found three types of warning responses: immediate evacuation, delay in evacuation, and refusal to evacuate for a wide range of reasons and rationales (Rahayu and Nasu 2011). Further investigation indicated that upon receiving both formal warning from the government and natural warning such as strong shaking, some people followed designated evacuation routes, while others acted spontaneously, even trespassing on their neighbors' properties (Rahayu 2012). Rahayu concluded that tsunami early warning can be achieved more effectively by understanding and accommodating the behavior of residents and using this information to plan or improve the downstream warning chain (Rahayu 2012). A functional tsunami early warning system requires timely and valid communication, information search and exchange directly among communities at risk, local-level SOPs, organized social networks, and community preparedness. Local communities face the greatest risk of tsunamis, yet they have the least resources for mitigating that risk (Spahn et al. 2014).

Establishing a functional tsunami early warning system in operational environments is a major, continuing challenge. Since 2011, Indonesia has served as a designated tsunami service provider, jointly with Australia and India, to the twenty-four nations that border the Indian Ocean. The IOTWMS has been tested in actual events and successfully issued a tsunami warning less than four minutes after an earthquake (Triyono 2012). However, the warning chain from detection instruments to the household level is long, tenuous, and vulnerable to disruption. BMKG has the authority to issue warnings only to the emergency operations centers of the provinces, cities, and villages. The direct transmission of warnings to the residents at risk depends on the policies, regulations, and actions of local authorities. The time, skills, and persistence needed to develop the funding, multifactor partnerships, and shared commitment to ensure a viable, sustainable warning system are often lacking at the local level, however (Thomalla et al. 2009).

For example, Padang City is one of the six Indonesian cities designated for investment in tsunami disaster risk reduction following the 2004 tsunami, and it has made substantial efforts to develop downstream tsunami warning system programs and plans, which were endorsed in local regulations in 2008 (Padang Mayor Regulation no.14/2008). Still, the city has not updated its warning program since 2010, nor has it considered the InaTEWS service guidelines issued in 2012. More difficult is that the city has not identified the "interface actors" between the city's EOC and the residents of the community at risk other than the local media stations. Especially under the urgent stress of an oncoming tsunami, designated responsible actors living in the community are very important and critical for rapid mobilization of residents to evacuate safely.

Reviewing studies for building resilience, other researchers have concluded that a community-based approach to building resilience is the most sustainable (Garcia and Fearnley 2012). That is, if community residents engage actively in the design and management of the system and participate in training exercises, they accept responsibility for the system and foster the local partnerships, volunteer activities, and leadership needed to make it operational in a sudden, urgent tsunami event (Rahayu et al. 2008; Spahn et al. 2014). Identifying the potential community leaders before a tsunami hazard occurs increases the level of cognition of risk in the community and enhances rapid communication, coordination, and control when an extreme event occurs (Comfort 2007). Incorporating a community-based, or people-centered, approach to tsunami mitigation policies is essential for effective implementation.

An Exploratory Study of Padang City: Design and Development of a People-Centered Tsunami Early Warning System

To address the complex problem of tsunami early warning and mitigation systems and to examine the design and practice of local communications processes, an exploratory study of local capacity and practice—the downstream component of an early tsunami warning system—was conducted in Padang City, an urban city on the western coast of Sumatra with high exposure to tsunami risk. The study was initiated in 2015 and continued in conjunction with other research activities through 2018. It stated three objectives:

1. To assess the gaps and missing links in the current policy regarding the tsunami early warning system, from end to end.
2. To identify potential actors in the community who could activate the warning network at the community level.
3. To assess the roles of the community actors and their potential for involvement in the community network.

Research questions

Several questions were outlined from the stated objectives of the study:

1. What are the gaps in the current early tsunami warning system?
2. Who are the potential actors at the community level for the tsunami early warning chain?
3. What are the roles and capacities of actors at the local level for the tsunami early warning chain?

Methods, data, and implementation

The study used a mix of quantitative and qualitative methods for data collection and analysis. First, document review of existing plans, policies, and procedures for tsunami warning and evacuation for Padang City was done to identify the gaps in the current early tsunami warning system (research question 1). These documents included Padang Mayor's Regulation 14/2010 regarding the Padang City tsunami early warning chain; Padang City's tsunami contingency plan; and the national tsunami early warning system SOP. Second, a focus group discussion (FGD) regarding the downstream tsunami early warning component was conducted on February 22, 2016, to address research question 2 regarding the potential actors at the community level and research question 3 regarding the roles and capacities of actors at the local level for activating tsunami early warning in practice. This FGD was attended by twenty officials from provincial and city agencies listed in Padang City's contingency plan. This discussion affirmed the importance of learning directly from both government and nongovernmental institutions regarding their views, responsibilities, and capabilities in the tsunami early warning process. Third, in parallel with the FGD, a survey of community residents in the Koto Tangah District of Padang City was conducted to cross-validate the FGD findings using a semistructured interview protocol for seventy-one respondents. The purpose was to gain insight into the expectations and past experiences of community residents' views regarding tsunami early warning messages and thereby create a profile of the community perspective of the tsunami early warning communications process. The survey used purposive and simple clustered sampling to increase the reliability of the data. Using these methods, thirty-six community leaders were identified among the seventy-one respondents as potential interface actors for early tsunami warning at the community level.

Data analysis

Three types of data were collected and analyzed using social network analysis (SNA), a method frequently used to identify the structure of relationships among groups and organizations engaged in emergency response operations, humanitarian assistance, and recovery after disaster events (Bisri 2016; Kapucu and Demiroz 2017; Kim and Hastak 2018). SNA can be used to identify quantitative measures of a network's structure, such as centrality, size, distance, and betweenness among the actors in a network. It can also be used to assess the qualitative context of the relationships among sets of actors and to identify subsets of actors within a larger network. In this study, SNA was used to identify the qualitative relationships between legal documents, formal rules, and informal emergent roles in the downstream component of the InaTEWS reported in the sources of data cited above were analyzed using UCINET, a software program that calculates network statistics and that also produces visualizations of the documented relationships (Borgatti et al. 2013).

The SNA networks described in table 2-1 served different purposes in the analysis. *Network 1* identified gaps and missing links in the existing InaTEWS program, based on legal documents and validated by findings from the FGD. *Network 2* verified the actual roles played by community leaders in practice as identified by participants in the FGD who represented government and nongovernment agencies with designated responsibilities listed in the legal documents. This network confirmed the midlevel actors in Padang who serve at the interface between the official national agencies transmitting early tsunami warnings and the community residents who need timely, valid information to guide their actions under threat. Participants in the FGD also raised questions that were included in the semi-structured interviews conducted in the survey of community residents at the district level. *Network 3* is based on findings from the survey of community residents at the district level and identified informal relationships that characterized both potential actors and vulnerabilities for community response to early tsunami warning. *Network 4* integrates networks 2 and 3 to propose a "whole community" framework for tsunami early detection and warning. Importantly, the set of networks evaluated the existing channels of information flow through the tsunami early warning chain and explored how the midlevel actors could improve the transmission, content, and speed of warnings to community residents

Results

Using UCINET6, a software program to calculate statistical measures of the relationships among the identified actors and information flow at downstream component, three types of data—data from legal documents, data from FGD findings, and data from the field survey—were analyzed to produce measures of centrality, betweenness, and visualizations of the four networks that characterize the community context for tsunami early warning in Padang City. Social network graphs represent the pattern of communication between the nodes (actor or organization) in an intended system. The findings were interpreted using additional insights and documentation from previous studies, document review, focus group discussion, and survey of district residents. The analysis illustrates the gaps and differences between the types of tsunami early warning networks based on the three types of data cited above.

TABLE 2-1. Networks identified for information flow in Padang City by data source and output, February 2016

Network Type	Data Input	Output
Network 1: Original network based on legal documents	Document review on existing plans regarding Tsunami Early Warning System (TEWS): National TEWS Standard Operating Procedures (SOP) Mayor regulation Tsunami contingency plan	Network matrix Identified list of actors to be invited for the focus group discussion (FGD)
Network 2: Government model developed based on FGD	The FGD was among the key stakeholders at the city-level TEWS conducted in February 2016 involved representatives from 20 identified organizations (governmental, nongovernmental, and private) from the provincial and city levels.	Approval by all stakeholders for a network regarding the city-level TEWS Identified local-level stakeholders who are potentially able to be involved in the people-centered Tsunami Early Warning Chain in Padang City
Network 3: Community model (people-centered) developed based on the result of study	Semi-structured interviews with 71 respondents from the community. The respondents were selected based on their subdistrict and roles in the community.	Current state of the downstream tsunami early warning system practice at the community level Identification of roles and capacity of the actors identified in the previous FGD

Network 1 (based on official documents). Padang City Mayor Regulation 14/2010 identified seven organizations and three communication channels to be included in the tsunami early warning chain for Padang City. These organizations were assigned formal responsibilities and are visualized in network 1. Table 2-2 presents a list of these organizations and their responsibilities. A network matrix based on the responsibilities specified in the regulation was created to produce a social network graph, as shown in figure 2-4.

As stated in the official documents, BMKG provides the tsunami warning to the Padang City EOC, called Pusdalops Kota, which then forwards the warning to all other organizations and institutions. Network 1, illustrated in figure 2-4, shows that the Padang City EOC (Pusdalops Kota) is the only organization in Padang City that receives the tsunami warning information directly from BMKG. This figure further shows that the city mayor (Walikota) receives the tsunami warning from the city EOC. The mayor makes the decision for an evacuation, based on the level of tsunami threat reported to the EOC. Figure 2-4 also illustrates the critical role of the Padang City EOC as a potential single point of failure in the communication line between BMKG as the national service provider and Padang City's mayor, which might result in a delay or even failure in issuing a timely evacuation order and mobilizing emergency response to an actual tsunami event.

City Mayor Regulation 14/2010 specified that the role of mayor of Padang City includes making decisions regarding actions to be taken after receiving the tsunami warning, such as issuing an evacuation order or public alert and implementing the incident command system. Thus, the Padang City EOC will not disseminate a tsunami warning and evacuation order before receiving the instruction from the mayor. This tie, presented as dark lines on figure 2-4, clearly depicts the redundant line of decision making and communication between the Padang EOC and the Padang mayor after receipt of the warning from BMKG and before it is disseminated to the community. This procedure may not be feasible and practical, as the tsunami might happen at any time, and would require the Padang City mayor to be available at all times. If the mayor has duties out of the city or abroad, this requirement would undermine or delay the tsunami warning process under this regulation.

Network 2 (government model, identified by participants in focus group discussion). Communication of tsunami early warning through the network, as identified during the FGD, has two main purposes. The first is to inform the last-mile community about an oncoming tsunami and order an evacuation. A second, identified by some organizations in the network, is to establish an incident command system during postdisaster emergency response. Similarly, this function was found in network 1 (original network) to be the responsibility of Padang City vice-mayor and secretary. These organizations are community health centers (*Puskesmas* and *Pustu*), police (*Kapolres*), public works agency (*Dinas PU*), development planning agency (*Bappeda*), social agency (*Dinas Sosial*), and several others. The second main purpose, reported from the FGD, is identifying other potential stakeholders who could serve as an interface between city agencies and the community. Two FGDs were conducted to gather information and insights directly from local officials and community residents. The first was conducted on February 26, 2016, and included representatives from

TABLE 2-2. Roles and responsibilities of actors as specified in the Padang City Mayor's Regulation 14/2010

Actor/Organization	Role and Responsibility	Node Symbol
Indonesian Agency for Meteorology, Climatology, and Geophysics (*BMKG Pusat*)	Provide tsunami warning information to Padang City emergency operations center (EOC)	⬤ (red)
Padang City emergency operations center/EOC (*Pusdalops Kota Padang*) and Disaster Management Agency (*BPBD/PK Kota Padang*)	– Receive warning information from BMKG Pusat – Propose suggestions for further actions to Padang City mayor – Relay warning information to the population at risk directly or through interface actors	◈ (red)
Padang City mayor (*Walikota Padang*)	Make decisions regarding actions to be taken following the tsunami warning	⧖ (red)
Padang City Communication and Information Agency (*Diskominfo–Kota Padang*)	Relay information from Padang City EOC to local television and radio channels	◈ (dark red)
Local television and Radio Channels (TVRI and Classy FM)	Relay information from Padang City EOC (Pusdalop Kota) and Communications and Information Agency (Diskominfo) to the population at risk through live broadcasts	◈ (yellow)
Indonesian Amateur Radio Union (RAPI)	Relay information from Padang City EOC (Pusdalop Kota) and Communications and Information Agency (Diskominfo) to the population at risk through live broadcasts	◈ (yellow)
Citizens band radio (RAPI)	Relay information from Padang City EOC (Pusdalop Kota) and Communications and Information Agency (Diskominfo) to the population at risk through live broadcasts	◈ (yellow)
Disaster preparedness community group (*Kelompok Siaga Bencana–KSB*)	Relay information from Padang City EOC to last-mile population in their respective communities	◈ (green)
Padang City vice-mayor (*Wakil Walikota Padang*)	Lead the preparation for emergency response based on tsunami early warning from Padang City EOC	▢ (dark red)
Padang City secretary (*Sekretaris Daerah*)	Assist preparation for emergency response based on tsunami early warning from Padang City EOC	▢ (dark red)

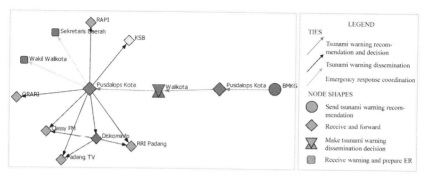

Figure 2-4. Network 1.

twenty organizations from Padang City and West Sumatra Province. The main discussion topic was how the tsunami early warning dissemination should be conducted in the city at the community level. The discussion produced a consensus for the ideal tsunami early warning transmission chain for Padang City, involving responsible organizations from the city and at the district and subdistrict levels. This information was incorporated into network 2, shown in figure 2-5.

The second FGD meeting was conducted three months later at the district level, the Padang Barat District. This one focused on identifying other potential stakeholders who could play interface roles between city agencies and the community. The roles of new midlevel actors—leaders from three nongovernmental organizations and two civil society organizations—were identified as serving important functions in the public dissemination of warning messages. These actors include the Red Cross (*PMI*), the tsunami preparedness community (*komunitas siaga tsunami/kogami*), Mercy Corps, the disaster preparedness school (*sekolah siaga bencana*), and the mosque community (*komunitas masjid*). The existence of these new midlevel-actors is further explored and discussed in network 3 (community model).

Based on these two FGD meetings conducted at city and district levels, network 2 (government model) was developed. In comparison to the original network, network 2 has more redundancy in the warning dissemination and involves more agencies at the city level. Moreover, critical ties such as links between the Padang City EOC, BMKG, and the mayor of Padang have been replaced by more robust backup communication ties. For example, in network 2, the regional office of BMKG, which is responsible for the West Sumatra region, has established several communication lines, such as with the Padang City mayor, several other

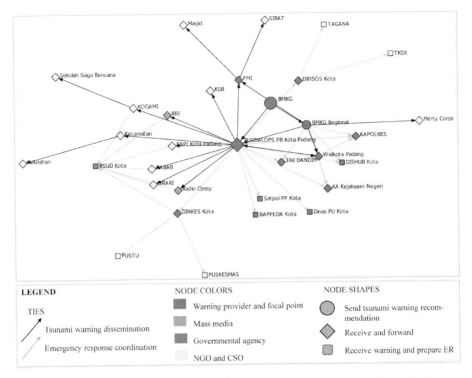

FIGURE 2-5. Network 2 (government model, based on focus group discussion).

agencies in Padang City (for emergency response functions), and several nongovernmental organizations and community leaders, for tsunami early warning dissemination.

Network 3 (community model, based on survey findings). To develop network 3 (community model), shown in figure 2-6, the potential roles of four types of community leaders as local interface actors were investigated: the sub-district office (*kelurahan*), disaster preparedness schools (*sekolah siaga bencana*), mosque communities (*masjid*), and disaster preparedness groups (*KSB*). Data were collected through an in-depth survey of seventy-one community residents using semistructured interviews in the Koto Tangah District, Padang City. Respondents were asked how they received or would receive evacuation orders in the case of tsunami. In addition, local interface actors were asked if they would communicate with other actors or convey the information to other community members. A network matrix of relationships reported by the respondents was created and entered into UCINET's visualization program to produce the diagram for network 3 as shown in figure 2-6.

According to the survey from the Koto Tangah District, various communication devices are involved in the dissemination of tsunami warnings to the last-mile community, households that do not have access to tsunami warning information or evacuation orders because of physical location or capacity. There are three main warning channels for tsunami warnings identified for the last-mile community: tsunami sirens activated by the Padang City EOC (32%), access to mainstream media such as television and radio channels (45%), and other telecommunication devices with their links or acquaintances in civil institutions (17%). (Multiple responses were accepted for the first three channels.) The remaining population must rely on warnings from local midlevel actors or community leaders (29%).

Discussion

Padang City is a coastal city with the highest risk of near-field tsunami in Indonesia. A tsunami hazard model shows a high probability of the occurrence of a Mentawai Megathrust tsunami to hit Padang City (Griffin et al. 2017). Since the 2004 Indian Ocean tsunami, disaster risk-reduction interventions have been implemented in Padang. It is also one of the six cities selected to model disaster risk reduction in Indonesia. Thus, the level of awareness and preparedness in the city toward tsunami risk is expected to be better than other coastal cities and regencies in Indonesia. Despite this expectation, study findings indicate several aspects that require improvement in the downstream warning chain of tsunami early warning in Padang City.

Under Indonesian disaster management law, UURI no. 24/2007, and in compliance with the autonomy of local governments in the Indonesian governmental system, the responsibility to protect the people from disaster lies at the local government level. The city mayor (or regent for regencies) has the highest responsibility to evacuate all the people at risk and leave no one behind. Upon receiving a warning from BMKG, and described in the InaTEWS grand scenario, the mayor is responsible for issuing an evacuation order (BMKG 2012).

As documented in the analysis of network 1, there were critical links of potential weakness in the Padang City tsunami early warning chain, including the sole reliance on the Padang City EOC to receive early warning information from the national service provider and the specified decision-making role of Padang City mayor. Although there is backup support from the regional office of BMKG West Sumatra for the Padang

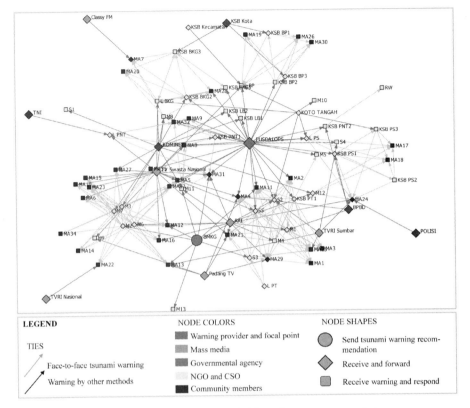

Figure 2-6. Network 3 (community model based on survey results).

City EOC, as reported in the FGD, these potentially weak links reflect the need to revise and update the city's regulation with advanced assessment of real-time field conditions. The existing regulation regarding the tsunami early warning chain in Padang was established in 2010, two years before the new guidelines of InaTEWS were released. The regulation is also a mayoral decree and has not yet been institutionalized in local law. This situation indicates the need for Padang City to renew or improve its regulation regarding the downstream tsunami early warning system.

Padang City's communication and information agency (*Diskominfo*) is the only institution with a role specified in the regulation but found to have no significant role in the practice. In the regulation, the agency has the responsibility of relaying information to mainstream media such as television and radio channels. In practice, these channels have already established informal ties with the Padang City EOC; the role of Padang City communication and information agency is only to ensure that the channels broadcast the information.

As a coastal city with the highest risk of a near-field tsunami, losses of minutes or seconds during the lead time for evacuation in Padang City could mean risking the loss of lives due to failed evacuation. These risks should be anticipated by simplifying the bureaucracy related to procedures in tsunami warning dissemination, as well as by increasing the redundancy in communication lines to strengthen the tsunami warning chain. As shown in network 1, assigning the sole decision-making responsibility to the mayor of Padang might slow down or undermine tsunami early warning and dissemination of an evacuation order.

The study's findings suggest that the early warning chain at the city level should be simplified. That could be done by mandating the predefined decision-making role to disseminate the warning and evacuation orders to the Padang City EOC so that the dissemination process could be initiated as soon as it receives the warning information, if necessary. In practice, the decision-making role of the Padang City mayor has been bypassed in most cases. It is feasible to reassign the decision-making role to the EOC. Supporting this need, the guidelines of InaTEWS have provided predefined actions to be taken by Indonesian city EOCs related to the level of tsunami threat: major warning, warning, and advisory.

The mayor of Padang does have a very critical and important role in exercising the authority to issue an order to mobilize evacuation during a tsunami emergency. In personal communication with community members in 2016, several community respondents reported that they had evacuated from a potential tsunami during the 2007 and 2009 Sumatra earthquakes after receiving a tsunami evacuation order conveyed by the incumbent Padang City mayor through the radio of the Republic of Indonesia, the local radio channel.

Although the Padang City EOC is the first and only gateway organization stated in the regulation as receiving warning information at the city level, in practice, the regional office of BMKG located in Padang Panjang could receive alerts for a potential tsunami or serve as a backup focal point in case the Padang City EOC faces communication difficulties. Documented in notes from the FGD meetings, the regional office of BMKG has established several communication ties with the Padang EOC, the mayor of Padang, and sectoral agencies in Padang City through a WhatsApp messaging group. Several other nongovernmental organizations and local community leaders were also identified as key actors to disseminate warnings to the public, covering some of the last-mile population.

Findings from this study conducted in the Koto Tangah District, Padang City served to cross-validate the findings from network 2 by showing the role of thirty-six midlevel actors and thirty-five active community members in network 3. Among the thirty-five community respondents, ten of them, or 29 percent of the study's sample of community members, could not receive tsunami warnings from any existing communication device, including sirens. Thus, network 3 has the capacity to disseminate the tsunami warning system to only 71 percent of the sample; that is, only 71 percent could receive tsunami warning messages, as shown in figure 2-7a. Meanwhile, the remaining 29 percent of the sample are categorized as the last mile, or people who do not have access to tsunami information through existing communication devices—for example, the siren that is operated by the Padang City EOC—in Padang City.

The community leaders (midlevel actors) identified above have their own capacity and network of communication. For example, the mosques community has a mosque network and can use the mosque speaker for alerting the surrounding community. Meanwhile, the disaster preparedness community groups have used a wide variety of communication networks, such as radio communication, WhatsApp messenger group, SMS, and even direct communications. The midlevel actors are the people in the community who have the first access to warning information and evacuation orders from the city EOC. The disaster preparedness schools use SMS and telephone to disseminate the tsunami warning information to parents and students. Finally, the subdistrict office uses walkie-talkies, WhatsApp messaging group, SMS, and the internet. Based on network 2 and network 3, an integrated model, network 4, was created as shown in figure 2-8. Walkie-talkies, also known as handie-talkies, are two-way radios that enable two-way verbal communication between two parties; each person may talk and listen to the other person in turn. These devices can be used within a distance of 0.5 to 2.5 kilometers without using pulse charges such as making calls.

Furthermore, the midlevel actors involved in the downstream tsunami early warning chain as interface actors have different characteristics, including varying levels of willingness to participate in the tsunami early warning chain as well as capacity to receive tsunami early warning information and to understand the warning information and evacuation order. This fact should be a consideration for further studies and

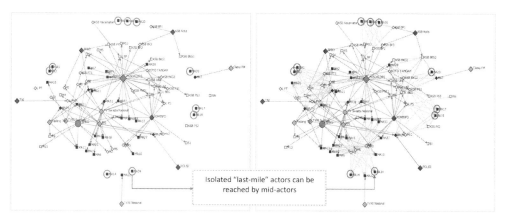

FIGURE 2-7. Comparison of Network 3 (a) without involvement of interface actors and (b) with involvement of interface actors.

planning for community-based early warning systems. As shown in network 3, several of the community leaders were already taking voluntary roles as interface actors. They were self-funded in purchasing walkie-talkie equipment to receive warning messages directly from the Padang City EOC. Some disaster preparedness group members who represent the last-mile population had been voluntarily attending city-level discussions and trainings regarding the tsunami early warning system.

Other findings of this study show that there are several flaws with the sirens installed in the city. A major problem is that only a small group of people really understand the meaning of each sound pattern of the sirens. Consequently, the intent of the siren patterns—such as Warning 1 (first warning message), Warning 2 (updated warning information), Warning 3 (tsunami has struck the coast), and Warning 4 (warning cancellation which means tsunami is over)—might not be fully conveyed to the last-mile community. Further, in personal communication with community members in 2016, several community members expressed their lack of trust in these sirens as these devices tend to be "unmanned" and "technologically dependent." In contrast, people might put more trust in their mosques' speakers. As described by Mulyasari and Shaw (2012), *masjids*, known as community mosques, are one form of religious civil society organization with relatively strong leadership by influential people from the community.

The use of the SNA method to identify communication patterns among these community members has found that community members

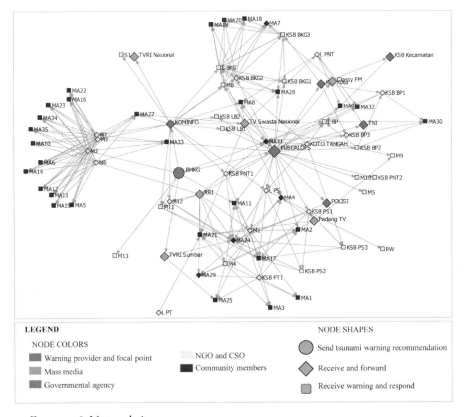

also established communication ties among themselves. Specifically, communication ties are identified between the disaster preparedness group members with the subdistrict officers and mosques community. The community leaders not only take the role of conveying information received from city level institutions, such as the Padang City EOC, but also share the information among themselves. This finding indicates that some partnerships are already established among these leaders, which need to be institutionalized.

The results and findings of this study have been communicated to the Padang City mayor and senior officials of several related institutions in Padang City (Rahayu et al. 2018) via a series of public engagement activities, such as a workshop on improving the downstream tsunami early warning SOP in Padang City, a tabletop exercise for the improved SOP in Padang City, and the 2016 Indian Ocean–wide wave exercise (IOWave16) to test the improved SOP. This IOWave16 end-to-end tsunami exercise not only successfully tested the performance of all stakeholders identified in network 4 of a people-centered early warning system, but also involved about thirteen hundred community members in an evacuation exercise to shelter as acknowledged and reported by ICG/IOTWMS IOWave16 task team in the IOWave16 exercise report (Intergovernmental Oceanographic Commission 2017).

After a series of public engagement events and testing during IO-Wave16, the improved SOP referred to as the people-centered tsunami early warning system for Padang City has been published as an academic paper on improving the Padang City Mayor Regulation 19/2018 regarding the tsunami early warning system.

Conclusion

The study discussed in this chapter identified possible improvements in the tsunami early warning system (downstream) for Padang City using an in-depth and holistic network model. Detailed data were collected through a series of FGD meetings with actors from the city and district levels and then combined with data obtained from semi-structured interviews in the field to recommend improvements in the downstream tsunami early warning chain (original network). The study produced a three-tier network model—that is, proposed network, enhanced network, and complete network—for a people-centered early warning system.

Four interface actors (midlevel actors) in the community—namely, subdistrict offices (*kelurahan*), schools (*sekolah siaga bencana*), community mosques (*masjid*), and disaster preparedness community groups—were identified that have different potential roles for their involvement in the downstream tsunami early warning chain, as well as different levels of willingness to participate and capacity to receive and disseminate the tsunami warnings to the public. To synchronize these differences, capacity building is necessary for these actors through training and provision of better communication devices to participate in the tsunami early warning chain.

The involvement and empowerment of community leaders have several advantages beyond the warning dissemination process. Community values, important in people-centered early warning systems, are identified in responses to the semi-structured interviews with community leaders and community members. Values such as volunteerism, leadership, and trust are highly advantageous to building capacity and may strengthen the people-centered early warning system among the communities. Moreover, the capacity of local police and the army (legally appointed as interface agencies in InaTEWS) might be hampered or overwhelmed by urgent demands during an emergency state and could benefit from the assistance of local community leaders. This likely situation further justifies the need to involve community leaders and focus on community-based tsunami early warning and preparedness. In building a solid partnership with the local government and the empowerment of these community leaders, these links of cooperation could not only be used for tsunami early warning chain in a near-field tsunami but could also be adapted for other types of sudden-onset hazards.

To conclude, this study contributed to improving the downstream early warning systems by developing recommendations for improving the people-centered tsunami early warning chain model for Indonesia. Several findings are presented as the strong points of this study:

- Removal of a single gateway agency or zero-redundancy in the downstream warning chain.
- A critical shift in the role of the mayor in the near-field tsunami warning system.
- Identification of a backup function of the regional office of BMKG to provide early information for a heads-up in the downstream warning chain.
- Finally, the identification of key interface actors (midlevel actors) in the community who have capacities to receive and relay warnings to the last-mile community.

This study offers an insight into the current gaps and challenges in the downstream tsunami warning system in Padang City. The results presented will contribute not only to improve the tsunami early warning chain of Padang City but also serve as guidelines for other coastal cities and regencies. The findings can be adopted as part of the existing tsunami early warning system, completing its end-to-end communications channel, by investigating its reliability and its compliance with the state of the art of InaTEWS. The development process for the three-tier network model can be replicated in other coastal cities and regencies in Indonesia, as well as in other regions that are exposed to near-field tsunami risk and have similar government systems that have autonomy at the local government level.

References

Basher, R. 2006. "Global early warning systems for natural hazards: Systematic and people-centred." *Philosophical Transactions of the Royal Society A: Mathematical, Physical and Engineering Sciences* 364, no. 1845: 2167–82. https://doi.org/10.1098/rsta.2006.1819.

Bisri, M. B. F. 2016. "Comparative study on inter-organizational cooperation in disaster situations and impact on humanitarian aid operations." *Journal of International Humanitarian Action* 1, no. 1: https://doi.org/10.1186/s41018-016-0008-6.

BMKG. 2012. *Tsunami Early Warning Service Guidebook for InaTEWS*. Jakarta: BMKG, Government of Indonesia.

BNPB. 2010. *2010–2014 National Disaster Management Plan*. Jakarta: BNPB, Government of Indonesia.

Borgatti, S. P., M. G. Everett, and J. C. Johnson. 2013. *Analyzing Social Networks*. Thousand Oaks, CA: SAGE.

Comfort, L. K. 2007. "Crisis management in hindsight: Cognition, communication, coordination, and control." *Public Administration Review* 67: 189–97.

Comfort, L. K. 2019. "Organizational Adaptation Under Stress: Tracing Communication Processes in Four California County Health Departments During the H1N1 Threat, April 28, 2009, to March 11, 2011." *American Review of Public Administration* 49, no. 2: 159–73.

Garcia, C., and C. J. Fearnley. 2012. "Evaluating critical links in early warning systems for natural hazards." *Environmental Hazards* 11, no. 2: 123–37. https://doi.org/10.1080/17477891.2011.609877.

Griffin, J. D., I. R. Pranantyo, W. Kongko, A. Haunan, R. Robiana, V. Miller, G. Davies, et al. 2017. "Assessing tsunami hazard using heterogeneous slip models in the Mentawai Islands, Indonesia." *Geological Society, London, Special Publications* 441, no. 1: 47–70.

Gunawan, E., P. Supendi, and T. Nishimura. 2020. "Identifying the most explainable fault ruptured of the 2018 Palu-Donggala earthquake in Indonesia using coulomb failure stress and geological field report." *Geodesy and Geodynamics* 11, no. 4: 252–57.

Intergovernmental Oceanographic Commission. 2017. *Exercise Indian Ocean Wave 16. An Indian Ocean-wide Tsunami Warning and Communications Exercise, 7–8 September 2016* (Vol. 2: Exercise Report No. 128; IOC Technical Series). UNESCO.

Kapucu, N. 2015. "Emergency management: Whole community approach." In *Encyclopedia of Public Administration and Public Policy*. https://doi.org/10.13140/2.1.2225.6965.

Kapucu, N., and F. Demiroz. 2017. "Interorganizational networks in disaster management." In *Social Network Analysis of Disaster Response, Recovery, and Adaptation*, ed. E. C. Jones and A. J. Faas, 25–39. Elsevier. https://doi.org/10.1016/B978-0-12-805196-2.00003-0.

Kim, J., and M. Hastak, M. 2018. "Social network analysis: Characteristics of online social networks after a disaster." *International Journal of Information Management* 38, no. 1: 86–96. https://doi.org/10.1016/j.ijinfomgt.2017.08.003.

Krackhardt, D. 1990. "Assessing the political landscape: Structure, cognition, and power in organizations." *Administrative Science Quarterly* 35, no. 2: 342. https://doi.org/10.2307/2393394.

Kumar, B. A., C. P. Kumar, E. P. R. Rao, P. L. N. Murty, V. C. Sekhar, K. S. Srinivas, M. V. Sunanda, J. Padmanabham, and S. S. C. Shenoi. 2019. *Source Constraints of 28 September 2018 Palu Tsunami from Simulations and Observations*. International Symposium and IORA Workshop on Lessons Learnt from the 2018 Tsunamis in Indonesia. Jakarta: BMKG.

Lindell, M. K., and R. W. Perry. 2012. "The protective action decision model: Theoretical modifications and additional evidence." *Risk Analysis* 32, no. 4: 616–32. https://doi.org/10.1111/j.1539-6924.2011.01647.x.

Luhmann, N. 1989/1996. *Ecological Communication*. Translated by John Bednarz. Chicago: University of Chicago Press.

Mawson, A. R. 2005. "Understanding mass panic and other collective responses to threat and disaster." *Psychiatry: Interpersonal and Biological Processes* 68, no. 2: 95–113. https://doi.org/10.1521/psyc.2005.68.2.95.

Mileti, D. S., and L. Peek. 2000. "The social psychology of public response to warnings of a nuclear power plant accident." *Journal of Hazardous Materials* 75, no. 2–3: 181–94. https://doi.org/10.1016/S0304-3894(00)00179-5.

Mulyasari, F., and R. Shaw. 2012, "Civil society organization and disaster risk reduction in Indonesia: Role of women, youth, and faith-based groups." In *Community, Environment and Disaster Risk Management*, ed. R. Shaw, 131–50. Bingley, UK: Emerald Group.

Natawidjaja, D. H., M. R. Daryono, G. Prasetya, P. L. Liu, N. D. Hananto, W. Konko, and S. Tawil. 2021. "The 2018 Mw7.5 Palu 'supershear' earthquake ruptures geological fault's multi segment separated by large bends: Results from integrating field measurements, LiDAR, swath bathymetry and seismic-reflection data." *Geophysical Journal International* 224, no. 2: 985–1002.

PuSGeN. 2019. *Report of Palu 7.4 M Earthquake, National Center for Earthquake Publication* (No. 1). PuS-GeN. https://sianipar17.files.wordpress.com/2019/07/laporan-kajian-gempa-palu-sulawesi-tengah.pdf.

Rahayu, H., R. Haigh, and D. Amaratunga. 2018. "Strategic challenges in development planning for Denpasar City and the coastal urban agglomeration of Sarbagita." *Procedia Engineering* 212: 1347–54. https://doi.org/10.1016/j.proeng.2018.01.174.

Rahayu, H. P. 2012. *Integrated Logic Model of Effective Tsunami Early Warning System*. Kochi, JP. Kochi University of Technology.

Rahayu, H. P. 2014. *IOTWMS Pillar 3: Tsunami Awareness and Response: Achievements, Challenges, Remaining Gaps and Policy Perspectives*. International Conference to Commemorate the 10th Anniversary of the Indian Ocean Tsunami, November 14, 2014. Jakarta: BMKG, Government of Indonesia.

Rahayu, H. P. 2019. *Disaster Literacy: Lesson Learned from 2018 Sunda Strait Tsunami*. International Symposium on the Lessons Learnt from the 2018 Tsunamis in Indonesia. Jakarta: BMKG, Government of Indonesia. https://unesdoc.unesco.org/ark:/48223/pf0000372721.locale=en.

Rahayu, H. P., and S. Kuraoka. 2019. "Strengthening the preparedness by transdisciplinary approach: Developing the guidelines for the tsunami early warning SOP and evacuation shelter in Indonesia." In *Transdisciplinary Approach (TDA) for Building Societal Resilience to Disasters—Concepts and Case Study for Practicing TDA*. TC21 ACECC (Asian Civil Engineering Coordinating Council) Publication.

Rahayu, H. P., and S. Nasu. 2011. "Logic model of people's mind toward tsunami early warning system." *Internet Journal of Society for Social Management Systems* 7, no. 1. http://ssms.jp/img/files/2019/04/sms11_4971.pdf.

Rahayu, H. P., In In Wahdiny, and Aria Mariany. 2008. *National Guideline for Tsunami Drill Implementation for City and Regency*. Jakarta: State Ministry of Research and Technology, Government of Indonesia.

Shannon, C. E. 1948. "A mathematical theory of communication." *Bell System Technical Journal* 27 (July and October): 379–423, 623–56.

Solihuddin, T., S., H., H., W., A., D., and D., P. 2019. *Rapid Assessment of the Sunda Strait Tsunami Impact in Pandeglang*. International Symposium on the Lessons Learnt from the 2018 Tsunamis in Indonesia. Jakarta: BMKG, Government of Indonesia. https://unesdoc.unesco.org/ark:/48223/pf0000372721.locale=en.

Sorensen, J. H. 2000. "Hazard warning systems: Review of 20 years of progress." *Natural Hazards Review 1*, no. 2: 119–25.

Spahn, H., M. Hoppe, A. Kodijat, I. Rafliana, B. Usdianto, and H. D. Vidiarina. 2014. "Walking the last mile: Contributions to the development of an end-to-end tsunami early warning system in Indonesia." In *Early Warning for Geological Disasters*, ed. F. Wenzel and J. Zschau, 179–206. Berlin: Springer. https://doi.org/10.1007/978-3-642-12233-0_10.

Taubenböck, H., N. Goseberg, N. Setiadi, G. Lämmel, F. Moder, M. Oczipka, H. Klüpfel, et al. 2009. "'Last-Mile' preparation for a potential disaster—interdisciplinary approach towards tsunami early warning and an evacuation information system for the coastal city of Padang,

Indonesia." *Natural Hazards and Earth System Science* 9, no. 4: 1509–28. https://doi.org/10.5194/nhess-9-1509-2009.

Thomalla, F., R. K. Larsen, F. Kanji, S. Naruchaikusol, C. Tepa, B. Ravesloot, and A. K. Ahmed. 2009. *From Knowledge to Action: Learning to Go the Last Mile.* Stockholm: Stockholm Environment Institute.

Triyono, R. 2012. *Current State of InaTEWS and Its Challenges.* UNESCO IOC TOWS Working Group Meeting, Paris.

Whittaker, J., B. McLennan, and J. Handmer. 2015. "A review of informal volunteerism in emergencies and disasters: Definition, opportunities and challenges." *International Journal of Disaster Risk Reduction* 13: 358–68. https://doi.org/10.1016/j.ijdrr.2015.07.010.

Wiener, N. 1948/1961. *Cybernetics or Control and Communication in the Animal and the Machine.* Cambridge, MA: MIT Press.

Chapter 3

A Reliable, Timely Communication Application to Enhance Tsunami Preparedness

Fuli Ai, X. Xerandy, Taieb Znati, Louise K. Comfort,
and Febrin Anas Ismail

The steadily increasing number of hazards over the past few years has led to a staggering number of escalating events, causing widespread damage to the environment and heavy losses in lives and property. Recent natural disasters in Indonesia underscore the gravity of such threats to coastal communities. The challenge of building resilience to hazards is recognized as a long-term effort to engage the "whole nation," including scientists, governmental agencies, private and nonprofit organizations, and communities (National Academies 2012). To meet this challenge, the flow of information needs to transform societal understanding of risk and enable self-organized, collective action to manage resilience of hazards at all levels. Decision making in disaster environments is best guided by a sociotechnical approach that integrates the science of the natural physical environment with analysis of the interdependent conditions among technical, organizational, cultural, and socioeconomic subsystems in communities at risk.

This study focuses on the increasing risk in coastal communities, specifically in Indonesia, and explores the process of building community resilience to hazards that involves geophysical, engineered, technical, computational, organizational, communication, and socioeconomic systems. Developing a scalable, disaster-tolerant, and socially aware information flow infrastructure as a core strategy for disaster management is critical to enhancing resiliency of coastal communities. The challenge is to embed technical components into the infrastructure communication and data dissemination processes that can capture the dynamics of the environment and adapt to interactions among physical, engineered, and sociotechnical systems that occur during hazard emergence and response within existing resource and time constraints.

Addressing this goal, we have designed a sociotechnical infrastructure to enhance downstream information flow in the Indonesian Tsunami Early Warning System (InaTEWS). This infrastructure includes the following:

1. A central coordinator (CC) unit that facilitates communication and collaboration among all stakeholders. The CC integrates services, monitors interaction among organizations and personnel, and transfers information among systems, enabling emergency managers in local governments to collect information from multiple sources, analyze local risks, and make decisions regarding response actions.
2. The Raspberry Pi network and satellite link to disseminate evacuation strategies and extend the communication network at community level.
3. A mobile geographic information system (GIS)–based smartphone application as social media.

Using this network of devices, community leaders assist local managers by monitoring environmental risk and informing residents of evacuation routes to the nearest shelter before the tsunami strikes. This study uses the city of Padang, West Sumatra, Indonesia, as a field site for investigating the technical infrastructure and social processes needed to enhance community resilience, as the dynamics of rapid adaptation and change for the whole community applies to other urgent hazards. This chapter translates the conceptual model of a bowtie network (see chapter 1) into a technical design for a neighborhood information infrastructure, building on the characterization of InaTEWS and risk assessment of Padang (see chapter 2). Further, it reports test findings from a field implementation of this sociotechnical network in Padang City.

The risk of disaster is particularly acute in Indonesia, which experienced four earthquakes above M_w = 7.0 (moment magnitude) scale in 2009 and reports earthquakes every year. Recent disasters in Indonesia—the 2004 Sumatran earthquake and tsunami; the 2009 Padang earthquake; the 2010 Mentawai Islands tsunami; the September 2018 Palu, Sulawesi, tsunami (*New York Times* 2018); and the December 2018 Sunda Strait volcano tsunami (*Guardian* 2018)—underscore the gravity of such threats to coastal communities.

The relentless increase in extreme events (United Nations 2010) compels a reexamination of models supporting the design, construction, policy development, and management of human communities and stresses the need for a more informed, integrated approach to understanding and managing hazards on a global scale. A National Academies report (2012) laid out the challenge of defining and building resilience to hazards as a long-term effort that would engage the "whole nation," including scientists, governmental agencies at all levels of jurisdiction, private and nonprofit organizations, and communities. To meet this challenge, *dynamic information processes must be in place to transform societal understanding of risk and enable self-organized, collective action to manage resilience of hazards at all levels.* In this context, decision making in disaster environments guided by a sociotechnical approach would integrate the science of the natural physical environment with analysis of the interdependent conditions among technical, organizational, cultural, and socioeconomic subsystems in communities at risk.

Although many definitions of resilience have been offered by the research community (Bruneau et al. 2003; Lentzos and Rose 2009; Comfort et al. 2010; White House and DHS 2011; National Academies 2012), we view resilience as a "web of practice" in communities that is supported by sociotechnical systems that adapt and change in interaction with one another and the environment. Designing community resilience requires understanding basic mechanisms of change in each of the component systems and identifying the interdependencies among them that shape the response of the community, positively or negatively, to hazard risk.

We focus on the increasing risk of hazards to coastal communities, specifically in Indonesia, and explore the process of building community resilience that involves recognizing the interdependent functions of geophysical, engineered, technical, computational, organizational, communication, and socioeconomic systems. The basic tenet of this work is that information technology (IT) plays a critical role in managing response to disaster. Damage to the IT infrastructure from extreme events further exacerbates the implementation of recovery efforts. Consequently, developing a *scalable, disaster-tolerant, and socially aware information flow infrastructure* as part of the core strategy in disaster management is critical to enhancing the resilience of coastal communities. We use the threat of *near-field tsunamis* (NFTs) in a location prone to this risk, Padang, West Sumatra, Indonesia, as a case study, as it includes all components of hazard evolution and hazard response. This hazard offers an excellent case for investigating the communications infrastructure and process to enhance community resilience, as the dynamics of rapid adaptation and change for the whole community are applicable to generalized hazard response operations that evolve at short timescales.

The structure of this chapter is as follows. First we present the theoretical framework for this research, complex adaptive systems of systems (CASoS) and its key properties (Glass et al. 2012). We then build on the current InaTEWS as presented in chapter 2 and propose a CASoS model of infrastructure in application to this case. Following that section, we explain the specific components of the bowtie design for the proposed information infrastructure to enable the community to become more adaptive and resilient.

Complex Adaptive Systems

We envision communities that learn to assess hazards endemic to the environment and that have the capacity to make collective decisions informed by scientific knowledge, leading to timely and effective risk reduction. Such communities operate *as ecological-socioeconomic-technical systems* or *complex adaptive systems of systems*. To manage diverse hazards and enhance resilience of the community at risk, a sociotechnical infrastructure should have the following properties.

Adaptiveness

The information and communication infrastructure, critical to disaster management, is distinctly vulnerable to a range of hazards. These events have varying potential for disruption depending on the type of event, its scale and location, and functionality of critical components of the communications infrastructure. To be adaptive and resilient, communications infrastructure needs to guarantee multiple communication channels for community members and government agencies. Accordingly, the infrastructure must incorporate different types of wired and wireless networks, including those that can be deployed on the fly, and leverage diverse paths offered by these networks to ensure sustainable, timely exchange of information (Comfort and Haase 2006).

Social awareness

The information and communications infrastructure must reflect and adhere to the cultural norms of the community at risk to enable effective and meaningful communications among its members. Social awareness, or social consciousness, has two meanings. First, it is the ability to recognize and understand the social context, problems, and constraints in which individuals routinely function in groups and communities. Second, it is the ability to sense and empathize with an individual's psychological state, thoughts, and feelings in relation to the specific social situation. These abilities enable individuals to build better connections with others, share information with more meaning, and create reciprocal interactions that support one another.

Social awareness leads to the development of social capital, that is, a reserve of *trust* among community residents that is a key component of self-organization in disaster environments. Social capital refers to connections among individuals—social networks and the norms of reciprocity and trustworthiness that arise from them (Putnam 2000). Basically, social capital is the degree to which individual members of a community share a set of common values and are willing to support one another voluntarily when needed (Kapucu 2011; Aldrich 2011). Community members need information provided by trusted entities to support decision making under threat. As emergency managers and community leaders work together to solve problems, they form trusted relationships and learn to support and rely on one another (FEMA 2011).

Community leaders elected by local governments and working in their communities have a greater capacity to recognize risk and form good relationships with the families and individuals they serve than do many other officials. They have already established trust within the community and can act as liaisons to open up communication channels. Information from trusted leaders is more likely to be received, understood, and accepted from these leaders than from other unknown sources or government agents (Mileti et al. 2002). Local governments can share tsunami warning and action strategies with neighborhood residents through community leaders to provide reliable information and support self-organizing action on a community-wide scale, increasing effective coordination (Andrulis et al. 2007).

Real-timeliness

A resilient community requires effective and timely information provided to operational personnel and residents in a condensed and actionable form to support decision making under threat. Timely, valid information is essential to conduct response activities and deliver assistance while minimizing delays and lost opportunities, yet it is difficult to capture a changing environment and exchange social information in real time.

NFTs represent a serious hazard in which decisions must be made by local governments immediately when the warning comes from national agencies, with limited stored data and an uncertain environment. Time is limited for threatened residents to reach safety. Social and traditional media are vital tools for sensing the community and providing information to the community; social media (Twitter, WhatsApp, and Facebook) users spread news rapidly in emerging crisis situations and can create viral information flows that may have marked organizational and social consequences. Social media have been used in disasters to mitigate harm and increase community resilience (Dufty 2012; Peary et al. 2012; Carley et al. 2016).

Interactivity

A people-centered, resilient community requires technological means to transmit information to carry out communication functions and inform community members regarding organizational processes; social networks facilitate the flow of communications and motivate resilient actions. It is vital for a local communication network to be multipronged, interconnected, and flexible to adapt to changing conditions. Technology could push information to, or pull information from, many community agents so that individuals may receive and confirm risk information shared within their social networks before deciding action (Wood et al. 2012). Otherwise, the information network should include a bottom-up feedback loop that allows residents to update timely environmental information to communities and local governments and enable community members to communicate with one another in the evacuation process. A resilient community is fundamentally social, reliant on interactions and relationships between and within communities and local governments (Paton et al. 2008).

We explore information and communication infrastructure to enhance community resilience to hazards, focusing on NFT risks in the Padang region. We refer briefly to the current tsunami early warning framework and systems used in Indonesia to disseminate information and organize evacuation during tsunamis, presented in chapter 2. We then discuss the limitations of the framework to enable effective community participation and improve its resilience to NFT hazards. Finally, we present an enhanced information infrastructure that harnesses the capabilities of the existing infrastructure and improves community resilience to NFT hazards.

Context: The InaTEWS framework

This study relies on the InaTEWS framework that was updated in 2019, as presented in chapter 2. As noted, the framework lacks a critical adaptive link between hierarchy nodes. Consequently, key actors may not have sufficient cognition about the real situation, as there are gaps in communicating changing risk conditions to community residents in real time. This lack of timely information exchange among key actors leads to less effective cooperation and coordination among agencies and community residents and may cause deadly delays in response to tsunami threats. In practice, the framework reveals the following areas of potential weakness.

Vulnerabilities of hierarchy in uncertain operational contexts

The analysis of the InaTEWS framework presented in chapter 2 notes a missing link between the local governments and the residents of coastal communities (see figure 2-2), a critical vulnerability in the system. This framework is subject to other vulnerabilities in the uncertain operational context of actual extreme events that need to be monitored and reinforced. Largely a *hierarchical, top-down, command-and-control–based* emergency management network (Jung and Song 2015), the InaTEWS framework can work with high efficiency in operations for which it is designed, but poorly in complex, dynamic environments of emergencies outside its specific context (Carlson and Doyle 2000). When any hierarchy's top nodes fail, the damaged structure isolates subnetworks from each other that are vital to the performance of the system (Carlson and Doyle 2000; Kapucu 2006), as shown in identifying the role of the mayor in the communications chain (see chapter 2). Hierarchy has key shortcomings in urgent environments that can be addressed through interactive, multiway communication processes at the community level.

Interaction

The InaTEWS infrastructure allows top-down and limited information flow in supporting two-way end-to-end communication. Only government agencies with designated responsibilities have authority to make decisions in disaster management, which discourages community involvement and social awareness. Decisions and the ultimate resilience of a community in practice are generally initiated by the actions of individuals and resources of communities at risk. Local residents are primary problem solvers and capable responders essential to improve the community's capacities for resilience (Mason 2006). Community resilience infrastructure is more than early warning technologies and standard response guidance; it provides informed strategies for community residents to use in protecting themselves during extreme events. The interdependency and interaction of community initiatives and government policy are critical for increasing resilience (National Academies 2012).

Coordination of roles and responsibilities

There are multiple levels of government in the InaTEWS system, and the roles and responsibilities for building resilience need to be effectively coordinated among them. In 2008, following the devastating 2004 Indian Ocean tsunami, the Indonesian government revised its organization of disaster management under Presidential Regulation No. 8. Primary responsibility for disaster mitigation, planning, and implementation was given to the National Disaster Management Agency (BNPD), but in the hierarchy of Indonesian governmental authorities, all subnational disaster management organizations are grouped under the Indonesian acronym BPBD, or regional disaster management agencies (Sec. 63, Presidential Regulation No. 8, 2008). This term, however, includes disaster management agencies at the provincial, regency, municipal, district, subdistrict, and sub-subdistrict levels in a complex chain of authority and responsibility intended to function under urgent, uncertain conditions of risk. The most challenging levels for tsunami mitigation and warning are the lower municipal/district/subdistrict/sub-subdistrict levels that represent the "last mile" of communication and warning regarding tsunami risk. This chapter focuses on developing the social and technical networks that enable timely mitigation and warning information to flow to the neighborhood residents at the lowest levels of governmental responsibility.

Not all local governments have a local disaster management agency (BPBD). Indonesia has currently established BPBDs in more than 90 percent of the districts/cities, but only 20 percent of the districts/cities have developed their risk assessment functions for lack of technical capacity, financial resources, and limited availability of detailed data at the district/city level. Only 25 percent of districts/cities have formulated hazard contingency plans (BNPB 2015), but the broad and complex task of community resilience relies on a collection of policies that need to be coordinated and integrated at multiple levels rather than framed as a single comprehensive governmental framework. Because emergency situations increase nonroutine tasks, community resilience needs to consider policy options and coordination across the full range of stakeholders and authorities that constitute the landscape of resilience. Identification of specific roles and responsibilities for government agencies in building resilience derives naturally from discussion of the complementary roles and actions that communities embrace as part of a systemic national effort to increase resilience.

Flexible authority

InaTEWS is the official tsunami early warning system in Indonesia. The National Tsunami Warning Center provides official data regarding undersea earthquakes or landslides to the Indonesian Meteorology, Climatology, and Geophysics Agency (BMKG) for scientific validation; BMKG is the only actor that has legal authority to issue tsunami early warnings (BMKG 2012). BNPB follows up on earthquake reports, tsunami early warnings, and advice issued by BMKG by transmitting threat information to communities and assisting them in preparing appropriate emergency response. BNPB does not have hierarchical authority over the provincial/regional/local BPBDs but supports them through provision of facilities and infrastructure, as well as technical assistance in many aspects of disaster management and disaster risk reduction. BPBDs address topics like conducting risk assessments, establishing local emergency centers and contingency plans, and implementing a comprehensive approach to promote local participation (BNPB 2013). Emergency situations create uncertainty, diverse conditions, and decreased centralization. Disaster is most effectively managed at the lowest jurisdictional level and supported by higher levels when needed. Decentralization means that it is not necessary that each level be overwhelmed prior to requesting resources from another level. Disasters begin and end locally, and most are largely managed at the local level. Many disasters in communities require unified response from local communities, some require additional support from neighboring or municipal/provincial jurisdictions, but only a few require support from national/federal jurisdictions (FEMA 2008).

Timely and consistent decision support

In Indonesia, budget restrictions resulted in many disaster risk-reduction activities being implemented by different stakeholders that were not based on the same framework. Implementation is sectoral in nature, not structured and integrated, which has made it difficult to allocate sufficient human, financial, or physical resources at all levels of government (BNPB 2009). The existing infrastructure does not provide the means to collect real-time information from the communities at risk to support decision making in government agencies. Policies with real-time information enable effective disaster preparedness and reduce losses (Allaire 2016), whereas policies with limited or untimely information from communities miss practicality and adaptability in uncertain, risk-prone situations. Community vulnerabilities differ even among communities with similar demographics and climate-related risks as social and political isolation inhibits access to sources of adaptive capacity (Hesed and Paolisso 2015). Inconsistency in information among government agencies creates inconsistency in government policies that may have unintended consequences and negatively impact community resilience.

Timely and adaptive communication to local residents

Communication and dissemination of disaster information are controlled by authority. Not all community members have direct access to disaster information. Lack of access to information concerning disaster risk is a major obstacle to community participation in disaster reduction, when information dissemination functions are not sufficiently developed in local communities. If the national agency issues a tsunami warning for a certain region and the warning reaches the intended government, but the local receiving agency does not have a functional warning dissemination system, there is no way to communicate the information to neighborhoods at risk. To reach the so-called last mile, adaptive communication means are essential for local governmental agencies to share timely risk information with community members who, in turn, share risk experiences with one another and the governmental agencies. This need was critically evident after the deadly 2018 Palu tsunami.

Socially aware information infrastructure for enhanced community resilience

In contrast to a hierarchical, top-down communications infrastructure for managing urgent events at the community level, researchers are developing frameworks to enhance risk awareness that include precautionary and preparedness measures to improve risk knowledge and share timely geographic information (Huang and Xiao 2015) and strengthen local communication and sensor networks (Roy-Chowdhury et al. 2015).

George and colleagues (2010) introduce wireless Ad Hoc and sensor network architecture, but this framework does not include visualization tools for the incident commander nor real-time actor tracking. The program developed by Heard and colleagues (2014) supports geospatial information, but this web-based system does not work when the communications infrastructure is degraded or the internet is down. Mohsin and colleagues (2016) have developed a program for use by first responders, like military and civilian organizations as a routine tool after disaster, but not for public use during disaster operations. Ganz and colleagues (2016) provide real-time visual analytics tools and a collaboration platform in different space and timescales. Dorasamy and colleagues (2017) propose a web-based knowledge management system to improve communication, coordination, and collaboration for government agencies interacting with vulnerable community and supporting agencies that are designed to share lifesaving knowledge and to support emergency management operations. These programs are promising, but none provide real-time, interactive communication at the neighborhood level in disaster-degraded environments.

In contrast, we investigate methods of assessing accurately and efficiently the dynamics of NFTs generated by undersea earthquakes or landslides as they impact communities at the neighborhood level. We design alternative strategies for monitoring, measuring, and communicating this risk to organizations and populations in coastal communities, model the actions that could be taken by policy makers and organizations in response to tsunami risk, and test a solution for measuring and communicating risk efficiently in an actual environment exposed to NFT hazards. This process is an iterative search for information under evolving conditions to inform decisions at multiple levels of action in response to the shared risk of NFTs.

Implementation of Bowtie Architecture

We use bowtie architecture as a model for information flow and decision support in risk conditions that require collective action (see chapters 1 and 6). The architecture identifies key sources of data that "fan in" simultaneously to a central processing unit, where the data are integrated, analyzed, and interpreted from the perspective and performance of the whole system. The new information is then "fanned out" to the relevant actors or operating units that use the information to make adjustments in their specific operations informed by the global perspective. This model fits well with an emergency operations center (EOC), where status reports from multiple agencies are transmitted to the service chiefs, who review the data from the perspective of the whole community. This architecture is depicted for two levels of the communication infrastructure: municipal and community networks. It recognizes the role that community residents play in risk management and extends the framework to enable self-organization and collective organizational actions by residents of the community, in collaboration with emergency personnel at the municipal level.

At the municipal/city BPBD, disaster managers (DMs) in the EOC receive tsunami alerts and earthquake information from the provincial/regency BPBD, forwarded from the national BNPB. With this information, they analyze reported risks for their specific municipality, design risk-reduction and mitigation strategies based on the available information and degree of uncertainty, and decide whether to disseminate evacuation warnings to the community. To upstream disaster managers—provincial/regency BPBD and national BNPB—municipal DMs send local data for integration, aggregation, visualization, and analysis. To downstream community residents, DMs disseminate earthquake information and tsunami early warnings. They also use real-time communication channels to provide instructive guidance to community members at risk, through neighborhood leaders and local institutions, to ensure that every resident in the threatened region receives the information in a prompt and timely manner. This *iterative* process allows information from one level of operations to support adjustments in performance in their specific environment, which in turn alter conditions at subsequent levels of operation in the system.

To support collective action at the neighborhood, or sub-subdistrict, level, residents of communities at risk are organized into networks of *leaders* and *followers*. Community leaders have valuable tacit knowledge about their neighborhoods and provide an in-depth understanding of the communities in which they live. They serve as critical links between official emergency response agencies and the residents they represent. Community leaders routinely organize emergency drills and public education efforts to raise risk awareness of community residents. They inform collective action with local hazard maps and evacuation plans and monitor environmental risk during tsunami evacuation. Followers accept direction from their leaders to evacuate collectively to designated shelters; they also communicate with their leaders to request help and

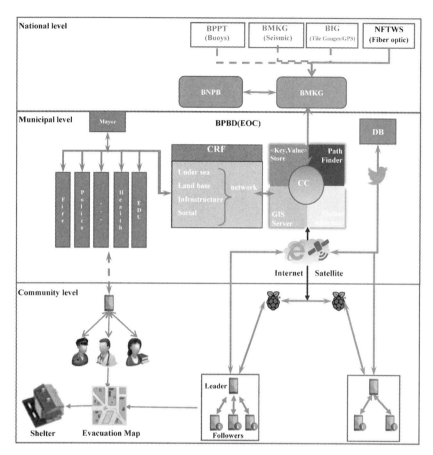

FIGURE 3-1. System framework.

report environmental risk. This dual role enables community residents to engage in collective action to reduce risk.

To support real-time interaction and communication between community residents and local DMs, we augment the existing tsunami early warning system with a cost-effective, sociotechnical infrastructure to enable real-time communication and interaction between multiple jurisdictions and community residents in an emergency situation. This infrastructure allows community members to share risk awareness—limited and life critical information. The proposed sociotechnical infrastructure, depicted in figure 3-1, includes a community resilience framework (CRF), a CC, a network of Raspberry Pis, and web-GIS–based smart application–driven mobile devices. The next section provides a detailed description of these components and discusses the capabilities and functionalities they provide to enhance community resilience to disaster.

Community Resilience Framework

The CRF is the dashboard central node of the bowtie architecture for a municipal EOC. It receives disaster risk data from multiple sources and integrates incoming data with existing data that characterize the community at risk. Incoming data are aggregated, indexed, and analyzed in reference to change from existing knowledge, and distributed according to the legal responsibilities of practicing agencies. The function of the CRF is to create a "common operating picture" of the community's changing status for action agencies. It creates a continuing information flow for real-time tsunami risk identification, assessment, understanding, and action.

The CRF includes two levels of information display. One level is designed for separate operational agencies—for example, fire, police, emergency medical services, and utility managers—to show the status of threat to the community for their specific responsibilities. This representation enables each agency to focus on its most urgent demands. The second level displays the status of the whole community so that DMs are able to identify risk and assess the impact of the threat on the entire community, design evacuation strategies, make decisions, and evaluate actions taken in relation to the safety of the whole community. The status of the community is updated as actions are taken in real time. Based on this GIS-based visual display of the changing status of risk, DMs can respond to an advancing threat and adaptive actions taken by other actors. Community leaders identified through the CRF are contacted to disseminate information, and the flow of communication among leaders and followers is monitored through the CRF. Incoming information from community leaders helps DMs at the EOC review and evaluate risk strategies continuously and develop or adjust risk management policies accordingly.

Central coordinator

The CC, the central processing unit of the CRF as shown in figure 3-1, coordinates communication and facilitates collaboration among all stakeholders. Real-time and stored data are simultaneously collected from different sources at the national, provincial/regency, municipal, district and subdistrict levels. These data are integrated, analyzed, and interpreted in the central unit, as it coordinates information from external and internal systems. The key task is to support safety evacuation path analysis and monitor community evacuation status; the models were described by Ai and colleagues (2016). For example, when the CC receives a "Show Me Safety" request from a community leader, it coordinates with the Shelter Allocator, Path Finder, and GIS Server models to return the safest path to the closest shelter that has sufficient capacity to host the leader and their followers. To fulfill this request, the CC invokes the Shelter Allocator model to select the closest available shelter and activates the Path Finder model to find the closest path to the selected shelter, based on the stored road network structure data, the real-time road network situation, and the GIS-based Location-Allocation Analysis model (ArcGIS Desktop 2013).

Raspberry Pi network and satellite link

To extend the communication network to the community level, we add a device called Raspberry Pi, an inexpensive single-board computer that is roughly the size of a credit card (Vujović and Maksimović 2015). This device is a flexible platform as it can operate in battery-powered mode. This feature is very useful in disaster operations, especially when the main power infrastructure may be damaged. This device can be configured and deployed to provide a temporary community-level multicast network structure (Ferdoush and Li 2014). To incorporate this device in multicast networks in practice, the network processing rate and energy consumption are significantly improved, and the network resource utilization can potentially reach its maximum instantaneously (Paramanathan et al. 2014). Raspberry Pi (Pi for short notation) devices may be attached to electric poles and trees along the road network. The Pi offers two types of wireless Wi-Fi networks. One type connects Pi devices to form a local network and requires the Pis to be set up close to one another to form an overlapping network that maintains stable connection. The other type serves as an access point for other mobile devices. The Pi can link to the Iridium satellite to create a nonoverlapping network topology that allows one Pi to connect with other Pis that are hundreds of meters away.

Pi networks provide communication infrastructure for communities to monitor, exchange, and disseminate information in real time. A mobile device is exposed to Wi-Fi when it enters the Pi wireless Wi-Fi coverage. The Pi network allows the device to communicate emergency information with other devices. When the device is moved across an overlapping Pi network, no disconnect occurs as it leaves the wireless coverage of one Pi and moves to the next, but there is a short-term disconnect and reconnect process in a nonoverlapping Pi network. During this time, the device will use the limited connect time to communicate. The satellite link offers an alternative channel to enable community residents to communicate with municipal government agencies when the internet is not functional. Further description of the Raspberry Pi network is presented in chapter 4.

Mobile device application

The criteria for mobile device applications designed for disaster environments are clear. People are less interested in information about the probability of disaster than in knowing what action they should take if a hazard occurs (Jones et al. 2008). They are more likely to follow a recommended strategy if it is easy to understand than if it is unclear (Wood et al. 2012). Information and messaging need to ensure effective communication with individuals who have disabilities, limited access to information, or functional needs, including being deaf, hard of hearing, blind, or of low vision (FEMA 2016). For wide adoption by community members, the application must be easy to operate, information must be in real time and trusted, and the interface must be clear and friendly. The mobile-GIS–based device application provides an emergency communication platform to operate between local EOC personnel and community leaders and among community leaders and their followers. The functions include (1) display of local basic map with road network, buildings, and other emergency facilities; (2) current GPS location of user; (3) receipt of tsunami warning; (4) evacuation path and timely notification of risk on map; (5) report of environmental risk; (6) message exchange with other members to request help and share risk information; and (7) analysis of evacuation path with offline network analysis model. The evacuation path is shown on the map with the current GPS location, familiar environment, and feature landmarks to enhance risk awareness during the time critical evacuation. Further description of communication features is presented in chapter 6.

Mobility tracking models

We developed models to track the dynamic mobility of users at municipal and neighborhood levels to ensure that emergency information could be accurately exchanged with users in real time. With mobility tracking models, the community leader can manage effective evacuation paths and report risk segments to find a safe detour to the designated shelter in a specific risk situation. Each user is assigned to a home Pi to track their real-time location in the Pi network during evacuation and ensure that a safe evacuation route can be sent to all residents of the neighborhood in time.

Governmental level. We use the key-value store (KVS) (Han et al. 2011) mechanism to track a community leader affected by a change in the environment. The KVS is a data storage and search function designed to store, retrieve, and manage associative data using a key that uniquely identifies a record and quickly finds relevant data within the database to match that record (https://www.gridgain.com/resources/glossary/key-value-store). For example, we design two datasets to match an evacuation path for the leader with segment condition models, respectively (Ai et al. 2016). First, the evacuation path model uses the unique road segment ID as a key to identify the leader whose evacuation path contains this segment. The value is composed of the leader's region ID, leader ID, walking direction, and segment order. This value points to a unique leader who uses this road segment, so the evacuation path is represented by a string of segment IDs; different leaders could share the same segment.

Second, the segment condition model only tracks the road segment at risk, but it also uses the segment ID as the key. The value is composed of the risk type that caused the road segment to be unavailable, like road congestion or blockage, building collapse, or bridge failure. The path information is displayed in both the local EOC server and the mobile device when the leader receives the evacuation path for the first time. During the evacuation phase, if a change affecting the road segment has been reported, the segment condition will be changed; the CC then uses the Segment ID as a key to notify all leaders affected by this segment and provide them with updated information to remedy the situation, such as a detour or direction to a safe shelter. The new path is then updated in the KVS.

Community level. We use a home agent to track the user's mobility. Each user is assigned to a specific Pi when registering in the Pi network; a care of address (CoA) is the temporary Pi for the user when moving into the coverage of a new Pi. This designation allows a home Pi to forward messages to the mobile device when the user is continually moving in the Pi network. When a mobile device moves to a new Pi, it sends a message to its home Pi to notify that it has moved into a new Pi coverage CoA. When another user or local government agent wants to communicate with the user, that person sends the information to the receiver's home Pi; then, the home Pi forwards the information to the CoA, including the sender's CoA. When the device receives the information, it replies directly to the sender's CoA. Meanwhile, a built-in GPS device in the smartphone records the actual personal evacuation path, which logs the user's geo-location (latitude and longitude) in a local file with a fixed time interval, for instance, every 2 seconds. The log file can be downloaded after the evacuation to analyze the movement patterns of single users and groups of users. Further description of the tracking scheme from the Raspberry Pi network perspective and mobility communication is found in chapters 4 and 7, respectively.

Community communication and interaction. This infrastructure provides a two-way, end-to-end real-time, dynamic interaction that includes both vertical and horizontal communications between municipal government personnel and community residents and among community residents. To function effectively, the infrastructure requires organizational design and preparedness training.

Effective preparedness enhances individual social awareness and improves emergency response. Prior to hazardous events, local tsunami preparedness training needs to be conducted for both local EOC personnel and residents. For example, the municipal EOC would manage the CRF and offer 24/7 tsunami warning service. Community residents would install the mobile device application in their own smartphones. Technicians would assist EOC personnel with extensive preloading of information to the community knowledge base that proactively reduces the traffic load during tsunami evacuation. Such information would include an offline map, community road network, and facilities network with shelters, hospitals, and schools. The application also includes a pre-coded communication template that is shared among all devices using the CRF, Raspberry Pi, and mobile device infrastructure.

The Raspberry Pi network with satellite link would be set up along streets that lead to safety shelters. Community candidate leaders would educate residents for disaster events and organize evacuation drills, managers in workplaces would train their employees, and teachers would train students in schools, giving special attention to "vulnerable groups" or those prone to damage, loss, suffering, injury and death in event of disaster (FEMA 2016). Emergency managers would prepare tsunami

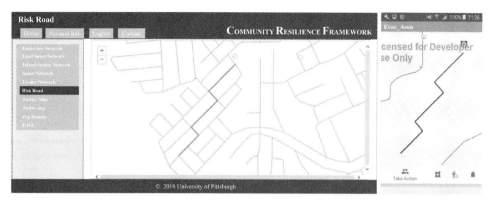

FIGURE 3-2. Evacuation route initialization.

contingency and evacuation plans, select potential evacuation shelters and other facilities, elect default activity leaders for each sub-subdistrict prior to a hazardous event, and identify default evacuation routes (see figure 3-2). Leaders use the mobile application to coordinate local resources, organize collective action, and report danger to local EOC personnel. Residents learn to find help, especially for vulnerable groups. Through preparedness training, neighborhood residents learn their roles and responsibilities, become aware of local tsunami risks, and become familiar with local warning and evacuation procedures. They are then ready to take action and follow their leaders when they receive a tsunami warning.

Government personnel to community residents

In general, there are three types of top-down hierarchical communication in a community at risk. The provincial/regional government sends a tsunami alert to municipal Emergency Operations Centers at risk. The local EOC sends a tsunami alert to the designated community leaders. Each community leader disseminates local evacuation information to all potentially affected neighborhood residents to enable them to act immediately. This networked strategy of communication ensures that all residents of the community can take timely, informed action. Details are presented below.

Tsunami alert. InaTEWS follows a top-down vertical information routine. When a municipal EOC receives a tsunami alert from regional-level governments via the national warning system, the DMs analyze the risk in terms of the tsunami's likely impact on their respective neighborhoods and populations, using the CRF. Based on multiple sources—including real-time information on population distribution and movement patterns, default evacuation paths and transportation network conditions, observational data on potential tsunami waves, technical communication networks using the internet, Raspberry Pi and satellite links, and condition of emergency facilities—the DMs make decisions to call for evacuation and provide guidance to communities at risk. They then disseminate evacuation commands via multiple communication channels, including text messages, the worldwide web, email, fax, public television, radio, sirens, mosque loudspeakers, and the smartphone application using the Raspberry Pi network and satellite link to community members in districts, subdistricts, and sub-subdistricts of the city (see figure 3-3). The whole evacuation process must finish before the NFT waves reach the shore, usually within minutes. Finally, the community leaders receive the safety warning forwarded from the local EOC and multicast it to their followers, who are already safe in the designated shelters.

Organize evacuation. The community leaders designated in the city's contingency plan assist residents in their respective neighborhoods by transmitting evacuation routes and self-organizing evacuation strategies to the nearest designated shelter before the tsunami arrives. Adaptive leaders are the link between the government and community. To maintain the communication link, an active leader needs to be available, so a neighborhood needs to elect several candidate leaders (Ai et al. 2016) and share the list with local government officials. The first candidate on the

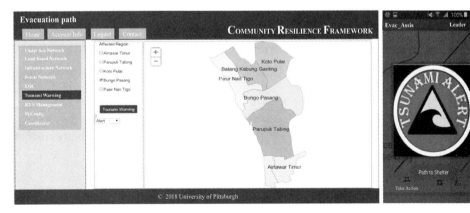

FIGURE 3-3. CRF sends tsunami alert to communities at risk.

list will be the default leader. When a tsunami alert reaches the community through the Raspberry Pi network, it triggers the election process for active candidate leaders to respond. If the default leader is available, that person will be the active leader; otherwise, the process will elect the next ranked leader, who will multicast the evacuation route to all followers in the neighborhood. The other candidate leaders are on reserve during the evacuation. Using this process, neighborhood residents collectively follow the evacuation route updated in real time (figure 3-4). Further description of the community leader election and small group first responder election in the Raspberry Pi network is described in chapter 7.

We investigate the latency on warning information streams from the local EOC to the community leader. The leader's location is measured based on the number of hops (one hop in routing is one transmission between two near nodes in the network) needed to traverse from the main Pi (Pi connected with the CRF) to the leader's current Pi. The EOC issues a set of warnings to community residents several times and logs the issuance time for each warning. The leader will acknowledge receipt of the warning back to the local EOC, and followers will acknowledge receipt back to the leader upon receiving the warning. For each notification, the local EOC and leader's devices will record the receipt time. The

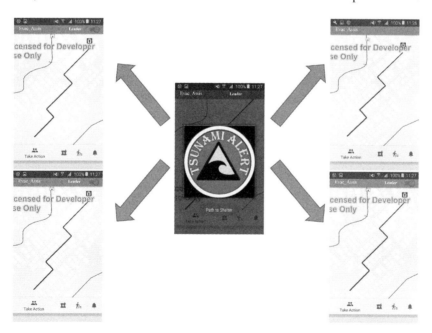

FIGURE 3-4. Leader disseminates evacuation route to followers.

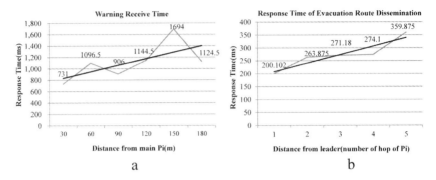

FIGURE 3-5. Tsunami warning and dissemination.

difference between these two timestamps implies the round-trip time of warning information flow.

Tsunami warning and dissemination. In internet mode, figure 3-5a shows the time needed for the information to reach the leader up to six hops, which imposes a distance of about 180 meters from the main Pi. The curve shows that in one hop, warning information can reach the leader in 800 milliseconds. The latency increases 100 milliseconds as the leader moves one hop away. It takes 1.4 seconds for the information to reach the leader, who is six hops away from main Pi. Once a leader receives the warning information, they disseminate the alert and default evacuation route to their followers. As figure 3-5b shows, it takes about 200 milliseconds for a follower to receive the alert from the leader when the follower is one hop distant from the leader. The latency increases around 350 milliseconds to reach a follower who is five hops away from the leader.

In satellite mode, we use Rock7 as the satellite communication provider. It uses the Iridium satellite network to provide this service. In this experiment, the main Pi unit connects to the satellite using a modem, and the main Pi connects to all other Pi units in its network. Meanwhile, the CRF connects to web service provided by Rock7 that bridges the communication between the Iridium satellite and the CRF. This configuration imposes several links to observe, including CRF–web service, web service–satellite network, satellite network–Pi network, and main Pi–mobile device. In this experiment, we test the infrastructure outdoors, where the modem is exposed to open sky with clear weather. Table 3-1 shows the latency incurred for each link. The time needed to send warning information to a leader's phone is 6 to 21 seconds. Compared to the

TABLE 3-1. Warning through satellite timeline

Case	Start Point	End Point	Duration (seconds)
1	CRF	Rock7 web service	< 1
2	Rock7 web service	Satellite network	6–21
3	Satellite network	Pi and mobile device	1.5
4	Mobile device	Satellite network	3–12
5	Satellite network	Rock7 web service	6–21
6	Rock7 web service	CRF	3–12

internet link, this configuration is slower. Nevertheless, it can be used as an alternative emergency communication channel if the internet link is broken.

Self-Organized Community

The entire community suffers from degraded communications after an earthquake. If the internet is down, communications capacity may drop sharply, and remaining capacity may be quickly depleted by immediate demands of affected residents. The adaptable technical infrastructure offers multiple ways to increase the communication capability for community members. It selects the best means of communication to enhance resiliency and coordinate activity. First, followers can communicate with followers, followers can communicate with leaders, and leaders can communicate with other leaders through the mobile devices. Over short distances, *horizontal* communication occurs between devices and does not require any infrastructure. Even in disaster conditions, community members can still communicate with one another. Second, in sub-sub-districts, community members can communicate through the internet or, if the internet is not available, could use a Raspberry Pi network. Third, if community members need to communicate over a wide distance with other communities or local governments in a disaster-degraded environment, they could communicate via the satellite link. Fourth, when community residents need help from local governments or need to report real-time environmental risks, they could communicate with upstream governments via satellite.

Preloaded message template

During evacuation, followers could communicate with their leaders to request emergency assistance and medical assistance through a Pi network

FIGURE 3-6. Follower finds medical assistance from leader.

(figure 3-6). To reduce the burden on communication infrastructure, avoid interference, and save critical time, especially when using satellite, a message template aids transmission. One packet size allows less than 50 bytes. The CRF, Raspberry Pi and smartphone share a message template and send minimum messages with necessary parameters. The receiver could fill the template, as the meaning of each button is coded and shared by the whole infrastructure. One-click sending is easy to operate and saves time, energy, and bandwidth during emergencies. For example, if a follower pushes the "need head trauma help?" button, message ID "1" will be sent to the leader in timely manner with minimum overhead.

When leaders receive a request, they will search the message based on its ID, and the message screen will show the follower's name, type, and emergency level. In the leader interface, there is a matched default button "head trauma help is coming" for quick response. When a follower needs special help not included on the default buttons list or normal message, a message could be typed manually.

Scalable infrastructure

The Pi network unit is organized for the sub-subdistrict level and could extend to the subdistrict and district levels. The geospatial boundary is clear, and the organizational boundary could adapt to the real-time evacuation situation. When a neighborhood leader receives a request for

TABLE 3-2. Message exchange timeline

| | Walking | | | Jogging | |
| | Total RTT (milliseconds) | | | Total RTT (milliseconds) | |
Hops	Average	Std Dev	Hops	Average	Std Dev
0	641.04	3,568.39	0	908.84	6,270.16
1	991.06	3,849.24	1	271.49	139.27
2	977.52	5,982.02	2	2,759.9	4,000.60

emergency help but cannot provide it, the leader may request assistance from other local leaders in adjacent neighborhoods. An available leader may respond and coordinate assistance to meet the request.

To test the information stream between community members, we developed a message exchange feature in the mobile device's application that is supported by the Pi network. In the test, the user moves around the network by walking and by jogging with a speed of about 1.7 meters per second and 3.75 meters per second, respectively. Transmission and receiving times are recorded for analysis. Once a user receives a message from another user, the recipient user needs to acknowledge it. In overlapping Pi coverage, the handover process takes about 15 milliseconds at the minimum and 28 milliseconds at the maximum while users move. The average round-trip time, which involves one transmission and its corresponding acknowledgment across different hops and mobility speed, is shown in table 3-2. Note that zero hop means that the users are in the same Pi.

Community Residents to Government Agencies

Access to information about causalities and damage is important to improve situational awareness and avoid invalid response during an emergency, for both DMs in governments and individuals needing assistance. Inaccurate and delayed information will force community residents to lose valuable time and incur increased risks. The prototype infrastructure provides a means to monitor and report the actual community situation to practicing DMs in real time. Community members act as trusted human sensors, monitoring the real-time environment and reporting local dangers to responsible community leaders. When three or more followers report the same danger or when the leader clarifies the risk, the report is forwarded to the municipal EOC. The reporting process is irreversible; it needs simple but strict authority controls. Only elected community leaders have authority to report a risk to local government officials and broadcast the risk to all nearby residents in the community (figure 3-7). Minimum information includes the location, type, and degree of the risk. Default risk buttons include road congestion, bridge collapse, building collapse, fire, and other types of disaster. The leader needs to double-check the report before submitting it to ensure that the information is accurate. If the reported road segment at risk is one part of the evacuation path, the report will trigger dynamic computation of a detour by the municipal CC and disseminate that message to affected leaders and followers.

To test the time duration of the reporting and detour process, we implement CC and ESRI ArcGIS Desktop with the Network Analyst extension module in Dell Precision Workstation M4800 as the server. The CC connects to the Raspberry Pi through a gigabit Ethernet link, and the leader and follower are in the same first main Pi. Table 3-3 shows the overall time from "leader send risk report" to "get detour" is 639.75 milliseconds. The CC needs about 576.5 milliseconds to compute the alternate detour itself. We consider this latency to be fast enough to support emergency operations.

Conclusion

We propose a social-technical infrastructure that has potential to enhance community resilience. When this infrastructure is used with the

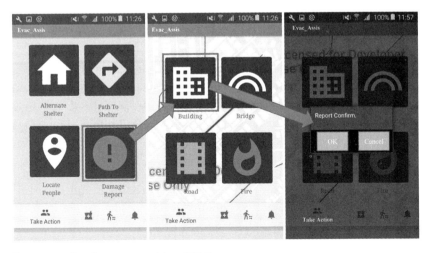

FIGURE 3-7. Leader reports risk to EOC.

Table 3-3. Risk report and detour timeline

Case	Start Time	End Time	Duration (milliseconds)
1	Leader sends risk report	Central coordinator (CC) receives report	31.625
2	CC receives report	CC computes detour	576.5
3	CC sends detour	Leader receives detour	31.625
4	Leader multicasts detour	Follower receives	70.08
Total	Leader sends risk report	Follower receives detour	709.83

current InaTEWS, it extends the electronic communication network to the community level. The warning, messaging, reporting, and detour processes augment the existing communications infrastructure with properties of adaptive, resilient, real-time, socially aware, and interactive functions.

This proposed communication infrastructure could motivate individuals, families, blocks, neighborhoods, and entire communities to develop updated collective action plans. It ensures timely information, prevents interference, reduces economic losses, and saves lives and trauma. The stakeholders will understand their roles and responsibilities clearly through real-time communication. Decision makers in local EOCs can increase their ability to receive tsunami alerts, recognize urgent threats, and dynamically adapt to disasters based on timely information from neighborhoods at risk. Through sharing information with others during the evacuation process, community residents enhance risk awareness. The social network can strengthen, and local knowledge and trusted relationships can amplify the power of communications and enhance emergency response.

Findings from this research, conducted in Indonesia where the risk of NFTs is high, will be applicable to coastal communities globally, including the West Coast of the United States. In the following chapters, we present rigorous tests of the connections among the set of electronic devices that make up the land-based network to ensure that this component is fully functioning, robust, and reliable to provide direct communication with neighborhood groups for operation in disaster-degraded environments. Further, we assess the extent to which the network of technical communication devices improves awareness, contextual understanding, and workflow among emergency response organizations and community residents, which ultimately contributes to coordinated response action and community resilience.

References

Ai, F., L. K. Comfort, Y. Dong, and T. Znati. 2016. "A dynamic decision support system based on geographical information and mobile social networks: A model for tsunami risk mitigation in Padang, Indonesia." *Safety Science* 90: 62–74.

Aldrich, D. P. 2011. "The power of people: Social capital's role in recovery from the 1995 Kobe earthquake." *Natural Hazards* 56, no. 3: 595–611.

Allaire, M. C. 2016. "Disaster loss and social media: Can online information increase flood resilience?" *Water Resources Research* 52, no. 9: 7408–23.

Andrulis, D. P., N. J. Siddiqui, and J. L. Gantner. 2007. Preparing racially and ethnically diverse communities for public health emergencies. *Health Affairs* 26, no. 5: 1269–79.

ArcGIS Desktop. 2013. "Location-allocation analysis." ArcGIS help library, ArcGIS 10.2, ESRI. Accessed October 13, 2020. http://desktop.arcgis.com/en/arcmap/latest/extensions/network-analyst/location-allocation.htm.

BMKG. 2012. "Tsunami Early Warning Service Guidebook for InaTEWS."

BNPB. 2009. "Indonesia National progress report on the implementation of the Hyogo Framework for Action (2007–2009)." Badan Nasional Penanggulangan Bencana, Jakarta. https://www.preventionweb.net/files/7486_Indonesia[1].pdf.

BNPB. 2013. "Indonesia National progress report on the implementation of the Hyogo Framework for Action (2011–2013)." Badan Nasional Penanggulangan Bencana, Jakarta. https://www.preventionweb.net/files/28912_idn_NationalHFAprogress_2011-13.pdf.

BNPB. 2015." Indonesia National progress report on the implementation of the Hyogo Framework for Action (2013–2015)." Badan Nasional Penanggulangan Bencana, Jakarta. https://www.preventionweb.net/files/41507_IDN_NationalHFAprogress_2013-15.pdf.

Bruneau, M., S. E. Chang, R. T. Eguchi, G. C. Lee, T. D. O'Rourke, A. M. Reinhorn, M. Shinozuka, et al. 2003. "A framework to quantitatively assess and enhance the seismic resilience of communities." *Earthquake Spectra* 19, no. 4: 733–52.

Carley, K. M., M. Malik, P. M. Landwehr, J. Pfeffer, and M. Kowalchuck. 2016. "Crowd sourcing disaster management: The complex nature of Twitter usage in Padang Indonesia." *Safety Science* 90: 48–61.

Carlson, J. M., and J. Doyle. 2000. "Highly optimized tolerance: Robustness and design in complex systems." *Physical Review Letters* 84: 2529–32.

Comfort, L. K., and T. W. Haase. 2006. "Communication, coherence, and collective action: The impact of Hurricane Katrina on communications infrastructure." *Public Works Management and Policy* 10, no. 4: 328–43.

Comfort, L. K., A. Boin, and C. C. Demchak, eds. 2010. *Designing Resilience: Preparing for Extreme Events*. Pittsburgh, PA: University of Pittsburgh Press.

Dorasamy, M., M. Raman, and M. Kaliannan. 2017. "Integrated community emergency management and awareness system: A knowledge management system for disaster support." *Technological Forecasting and Social Change* 121: 139–67.

Dufty, N. 2012. "Using social media to build community disaster resilience." *Australian Journal of Emergency Management* 27, no. 1: 40.

FEMA (Federal Emergency Management Agency). 2008. "National Response Framework." https://www.fema.gov/pdf/emergency/nrf/nrf-core.pdf.

FEMA. 2011. "A whole community approach to emergency management: Principles, themes, and pathways for action." Washington, DC: Federal Emergency Management Agency.

FEMA. 2016. "National Mitigation Framework." https://www.fema.gov/national-mitigation-framework.

Ferdoush, S., and X. Li. 2014. "Wireless sensor network system design using Raspberry Pi and Arduino for environmental monitoring applications." *Procedia Computer Science* 34: 103–10.

Ganz, A., J. M. Schafer, Z. Yang, J. Yi, G. Lord, and G. Ciottone. 2016. "Evaluation of a scalable information analytics system for enhanced situational awareness in mass casualty events." *International Journal of Telemedicine and Applications*. https://doi.org/10.1155/2016/9362067.

George, S. M., W. Zhou, H. Chenji, M. Won, Y. O. Lee, A. Pazarloglou, R. Stoleru, and P. Barooah. 2010. "DistressNet: A wireless ad hoc and sensor network architecture for situation management in disaster response." *IEEE Communications Magazine* 48, no. 3: 128–36.

Glass, R. J., W. E. Beyeler, A. L. Ames, T. J. Brown, S. L. Maffitt, N. Brodsky, P. D. Finley, T. Moore, M. Mitchell, and J. M. Linebarger. 2012. *Complex Adaptive Systems of Systems (CASoS) Engineering and Foundations for Global Design.* No. SAND2012-0675 (January). Albuquerque, NM: Sandia National Laboratories.

Guardian. 2018. "Sunda strait tsunami volcano Indonesia." December 24, 2018. https://www .theguardian.com/world/2018/dec/24/sunda-strait-tsunami-volcano-indonesia.

Han, J., E. Haihong, G. Le, and J. Du. 2011. "Survey on NoSQL database." In *Proceedings of the 6th International Conference on Pervasive Computing and Applications (IPCA)*, 363–66. New York: IEEE.

Heard, J., S. Thakur, J. Losego, and K. Galluppi. 2014. "Big board: Teleconferencing over maps for shared situational awareness." *Computer Supported Cooperative Work (CSCW)* 23, no. 1: 51–74.

Hesed, C. D. M., and M. Paolisso. 2015. "Cultural knowledge and local vulnerability in African American communities." *Nature Climate Change* 5, no. 7: 683.

Huang, Q., and Y. Xiao. 2015. "Geographic situational awareness: Mining tweets for disaster preparedness, emergency response, impact, and recovery." *ISPRS International Journal of Geo-Information* 4, no. 3: 1549–68.

Jung, K., and M. Song. 2015. "Linking emergency management networks to disaster resilience: Bonding and bridging strategy in hierarchical or horizontal collaboration networks." *Quality and Quantity* 49, no. 4: 1465–83.

Kapucu, N. 2006. "Interagency communication networks during emergencies: Boundary spanners in multiagency coordination." *American Review of Public Administration* 36, no. 2: 207–25.

Kapucu, N. 2011. "Social capital and civic engagement." *International Journal of Social Inquiry* 4, no. 1: 23–43.

Lentzos, F., and N. Rose. 2009. "Governing insecurity: Contingency planning, protection, resilience." *Economy and Society* 38, no. 2: 230–54.

Mason, Byron, ed. 2006. *Community Disaster Resilience: A Summary of the March 20, 2006 Workshop of Disasters Roundtable.* Washington, DC: National Academies Press.

Mileti, D. S., L. Peek, and P. Stern. 2002. "Understanding individual and social characteristics in the promotion of household disaster preparedness." In *New Tools for Environmental Protection: Education, Information, and Voluntary Measures*, 127–32. Washington, DC: National Academies Press.

Mohsin, B., F. Steinhäusler, P. Madl, and M. Kiefel. 2016. "An innovative system to enhance situational awareness in disaster response." *Journal of Homeland Security and Emergency Management* 13, no. 3: 301–27.

National Academies. 2012. *Disaster resilience: A national imperative.* Committee on Increasing National Resilience to Hazards and Disasters and Committee on Science, Engineering, and Public Policy. Washington, DC: National Academies Press.

New York Times. 2018. "Indonesia tsunami Sulawesi Palu." September 29, 2018. https://www.nytimes .com/2018/09/29/world/asia/indonesia-tsunami-sulawesi-palu.html.

Paramanathan, A., P. Pahlevani, S. Thorsteinsson, M. Hundeboll, D. E. Lucani, and F. H. Fitzek. 2014. "Sharing the pi: Testbed description and performance evaluation of network coding on the Raspberry Pi." In *2014 IEEE 79th Vehicular Technology Conference (VTC Spring)*, 1–5. New York: IEEE.

Paton, D., B. Parkes, M. Daly, and L. Smith, L. 2008. "Fighting the flu: Developing sustained community resilience and preparedness." *Health Promotion Practice* 9, no. 4 (suppl.): 45S–53S.

Peary, B. D., R. Shaw, and Y. Takeuchi. 2012. "Utilization of social media in the east Japan earthquake and tsunami and its effectiveness." *Journal of Natural Disaster Science* 34, no. 1: 3–18.

Presidential Regulation No. 8. 2008. Presidential Regulation of the Republic of Indonesia No. 8/2008 concerning the National Agency for Disaster Management: Sec. 63. Jakarta, Indonesia. https://www.informea.org/en/legislation/presidential-regulation-republic-indonesia

-no-82008-concerning-national-agency-disaster#:~:text=Presidential%20Regulation
%20of%20the%20Republic%20of%20Indonesia%20No.,working%20arrangements%20of
%20the%20National%20Disaster%20Management%20Agency.

Putnam, R. D. 2000. *Bowling Alone: The Collapse and Revival of American Community.* New York: Simon and Schuster.

Roy-Chowdhury, A. K., M. Kankanhalli, J. Konrad, C. Micheloni, and P. Varshney. 2015. "Introduction to the issue on signal processing for situational awareness from networked sensors and social media." *IEEE Journal of Selected Topics in Signal Processing* 9, no. 2: 201–3.

United Nations. 2010. "Natural hazards, unnatural disasters: The economics of effective prevention." World Bank. November 15, 2010. http://www.worldbank.org/en/news/feature/2010/11/15/natural-hazards-unnatural-disasters-the-economics-of-effective-prevention.

Jones, L., R. Bernknopf, D. Cox, J. Goltz, K. Hudnut, D. Mileti, S. Perry, et al. 2008. "The Shake-Out Scenario: U.S. Geological Survey Open File Report 2008-1150 and California Geological Survey Preliminary Report 25." US Geological Survey. https://pubs.usgs.gov/of/2008/1150/of2008-1150small.pdf.

Vujović, V., and M. Maksimović. 2015. "Raspberry Pi as a sensor web node for home automation." *Computers and Electrical Engineering* 44: 153–71.

White House and DHS (The White House and the Department of Homeland Security). 2011. "Presidential Policy Directive-8." http://www.dhs.gov/xlibrary/assets/presidential-policy-directive-8-nationalpreparedness.pdf.

Wood, M. M., D. S. Mileti, M. Kano, M. M. Kelley, R. Regan, and L. B. Bourque. 2012. "Communicating actionable risk for terrorism and other hazards." *Risk Analysis: An International Journal* 32, no. 4: 601–15.

Chapter 4

Device-to-Device Communication: A Scalable, Socially Aware, Land-Based Infrastructure to Support Community Resilience in Disaster Events

X. Xerandy, Fuli Ai, Taieb Znati, Louise K. Comfort, and Febrin Anas Ismail

In general, resilience can be viewed as capacity to cope with adverse events. In this study, we define resilience, in concurrence with the National Research Council (2012, 16), as the *ability to prepare, absorb, recover from, and more successfully adapt to adverse events*. In the context of disaster management, the Federal Emergency Management Agency (2011) introduced a new paradigm of building community resilience to disaster that involves the community's active participation and collaboration rather than relying on a government-centric paradigm. Engaging the "whole nation," including government organizations at all levels of jurisdiction, is essential to build community resilience to disaster in the long term. Today, this paradigm is being implemented and tested in other nations as a means to strengthen communities against recurring hazards.

Designing community resilience to hazards requires understanding the basic mechanisms of change in each of the components of interacting sociotechnical systems and identifying the interdependencies among them that shape the response of the community, positively or negatively, to hazards (Comfort 2020). Community and government capacity to manage and understand the risk associated with hazards depends on the ability to learn and adapt collectively to changes before, during, and after disaster events. Ensuring information exchange and knowledge sharing becomes the key factor in maintaining a community's capacity to adapt to disaster-prone environments. In an actual disaster event, characterized by uncertainties and rapid change, lack of timely information search and exchange can hinder proper action and decision-making processes at the community and organizational levels of operation (Nowell and Steelman 2015; Comfort 2019).

Communication in Disaster Events

Involving community and government actors as integral stakeholders in building resilience to hazards requires functional communications among them. During extreme events, when uncertainties unfold due to abrupt changes that occur in the environment physically or socially, communication between community residents and government actors as first responders in the disaster event may be severely disrupted (Comfort and Haase 2006).

Communication during a disaster emergency includes warning messages, evacuation directives, response status, available assistance, and other protective actions (Klein et al. 1993). However, rapid, dynamic environmental changes during the disaster emergency also may lead community actors to react differently upon receiving information compared to day-to-day messages. Information accuracy, timeliness, presentation, and source and context-relevance, combined with the individual recipient's education level, age, and culture, play critical roles for a community to understand and act accordingly (Comfort 2007). This capability is even more affected when information conveys the difference between life and death.

From the perspective of local government authorities, effective communication across multiple agencies during disaster emergency is essential to conduct effective and comprehensive evacuation and operations during and after disaster events (Kapucu 2006). This ideal collaboration is difficult to achieve in practice, especially during the urgent stress of an emergency. Differences in organization structure and personnel skill, compounded by different standards of communication tools and damaged existing infrastructure, have hindered the process.

Research Overview

In disaster events, the impacted community often suffers from limited communication access to get critical information and request further assistance (Mileti 1999). In addition to the cross-agents' coordination and communication problem, local authorities and emergency responders also need to disseminate critical information selectively during the disaster event. All information should be exchanged to the right recipients in a timely manner. Distributing information to the wrong place and at the wrong time could result in an uncontrolled, chaotic situation in the community, which impedes disaster operations.

Hazards can strike any country or community; hence, building resilience becomes a compelling need (National Research Council 2012). Building collective community resilience to hazards is not an inexpensive task. For newly developing countries, cost can be a crucial consideration in carrying out such approach and should be taken into account in building communications infrastructure (Xerandy et al. 2016).

Clearly, strengthening information streams across the multiple components of a sociotechnical system would be the core strategy to enhance community and government capacity and knowledge to act collectively and self-organize to reduce hazards and respond to extreme events (Comfort et al. 2006). Information infrastructure must be in place to ensure a timely and reliable information and communication process. These concerns bring two research problems to address. First, what are the essential properties that the information infrastructure should have to support the underlying communications process, particularly in the disaster emergence context? Second, what type of infrastructure can meet those expected properties?

Approach

With the goal of engaging community residents to act collectively in reducing risk from hazards, assuring information and knowledge sharing among disaster management personnel and mitigation actors holds a crucial key for more effective risk management of extreme events (Comfort and Zhang 2020). Specifically, supporting these processes could improve evacuation during an actual emergency event, as it can give invaluable incentives for more informed community action and emergency services. As addressed by Arneson and colleagues (2017), information deficit during disaster emergence can affect the evacuation operation.

Concerning the importance of the information exchange and knowledge-sharing process during disaster emergence when rapid changes of environment may hinder the process, we define the following five properties that an information infrastructure should have.

1. *Timeliness.* In disaster operations, time is a critical factor; the infrastructure must be able to deliver information within time constraints delineated by the underlying application. Emergency operations in disaster contexts are planned, organized, and executed within a strict timeline. Timely information delivery heavily determines the success of the operations (Klein et al. 1993). This function is critical in disaster early warning, as the warning must

be disseminated to community residents at risk in time to mitigate disaster impact.

2. *Reliability.* The infrastructure should be able to support information delivery to the affected community in unfavorable environmental conditions before, during, and after natural hazard events (Ai et al. 2016). This concern entails the need of multiple paths for communication channels. Providing alternate paths of communication will prevent information delivery failure (Fountain 2001; Graber 2002). Having reliable information infrastructure also imposes robust design of information-sharing applications that run on top of the infrastructure so that it would not easily crash during disaster.

3. *Social awareness.* With the community as the primary actors in this paradigm, the infrastructure should be designed to conform with the social structure, norms, and values of the community of interest (Tierney 2014). This socially aware design principle will allow more effective community participation and self-organizing.

4. *Cost-effectiveness.* The implementation cost must be minimized without penalizing overall system performance significantly (Xerandy et al. 2016.) Furthermore, the infrastructure should be scalable in its implementation.

5. *Robustness.* Despite being developed as cost-effective infrastructure, the infrastructure is still required to operate in a degraded environment. This property requires the design of the infrastructure that can withstand dynamic environment changes (Bruneau et al. 2003). The infrastructure should also be easy to deploy. Using simple, yet advanced, disaster-tolerant components will improve infrastructure robustness.

Allowing diverse information flows and knowledge within the community and across agencies to local authorities without being focused and managed could result in counterproductive outcomes. The bowtie conceptual framework (Comfort 2005; Comfort et al. 2009) addresses this issue by managing how these diverse information flows should be collected from different levels of agencies and sources, analyzed, visualized, interpreted, and distributed to relevant recipients and decision makers to support more informed actions. We propose a timely, reliable, socially aware, cost-effective infrastructure to meet the stated criteria for a neighborhood-level communications framework

To meet those properties, this study leverages advanced communications and networking technology as the core tool. We propose a land-based communications infrastructure that exploits the advantages of a network of single-board computers (SBCs), coupled with mobile phones and a centralized, Java-based web service featuring visual displays through geographical information system (GIS) technology. Specifically, this chapter presents the design and development of a network of SBCs using wireless Raspberry Pi devices to bridge critical information exchanges among community residents and between community residents and local government personnel.

In addition to the network of SBCs, this chapter discusses the use of opportunistic communication, as an emerging ad hoc, peer-to-peer wireless communication technology, to serve as a viable tool that enables the spontaneous formation of self-organized citizen networks for collaborative action. More specifically, we develop a device-to-device (D2D) communication scheme using mobile phones as a virtual augmentation of the proposed SBC network to provide more communication access to a community at risk. This communication scheme is independent from existing, regular infrastructure communication infrastructure such as cellular technology. To have such emergency communication means that it can operate without support from such infrastructures and could offer immense benefits in a disaster event, particularly in areas with severely damaged infrastructure due to the disaster.

To demonstrate the socially aware design aspect in our network, we designed the network based on the actual community social structure. We selected Padang City, West Sumatra, Indonesia, as the case study. This city is located on the west coast of Sumatra Island and has the highest population in the region. Based on the study in McCloskey and colleagues (2007), the marine region encompassing the western coastline of Sumatra Island poses a potential epicenter that could generate a near-field tsunami threat to Padang that could strike the city in less than one hour (Sasorova et al. 2008). With this imminent threat of tsunami, the government at the local, as well as the national, level actively seeks to enhance the community's capacity to mitigate the risk of tsunami disaster.

As part of a community empowerment program to encourage active participation for disaster mitigation, the Padang City government has organized the community into smaller designated groups (known as KSBs) in which each group has a leader. Instead of being officially appointed by government, the leader rises from within the community itself, a trustworthy person acknowledged by that community. In the context of this application, local government and emergency services consider the leader to be the point of contact for the neighborhood in case of an emergency. In any region, one or more leaders may be present. The leader is responsible for managing, organizing, and leading the community in collective action during evacuation. Consequently, the leader needs to be physically and mentally fit.

This social structure makes Padang City an appropriate case study for our proposed network. The neighborhood leader is regarded as the default leader when the evacuation process is in progress and is the responsible point of contact for the group. To increase the robustness of alert dissemination within this social structure, backup leaders may also be needed. The backup leader will be promoted to leader in charge when the original leader cannot be reached. The scheme of promoting this leader is described in detail in chapter 7.

Related research

Addressing the need for timely, accurate communication regarding tsunami warning and evacuation requires a comprehensive, integrated approach that can accommodate diverse information and communication technologies and platforms and can operate before, during, and after a disaster event. A broad range of information and communication technologies—including wireless mobile technology, web service, social media, remote sensing, GIS, satellite communication, and sensors—have been exploited to enhance communication quality in disaster events. In some cases, a combination of different technologies and platforms are sought to provide a more comprehensive solution (APCICT 2011).

Integrating existing communication infrastructures, combined with layer-based reinforcement and cloud processing, has been proposed by Ali and colleagues (2015). With this approach, the authors designed a communication network architecture that allows for reliable, robust, and timely communication and that can be deployed within reasonably short time, which is particularly important in an adverse event.

Meanwhile, Bjerge and colleagues (2016) discuss the potential of online platforms to gain unique insights into the coordination behavior among disaster management agencies and individual actors. Relevant to this case study, the authors developed a virtual on-site operation and coordination center. This platform, developed under the administration of the United Nations Office of Humanitarian Affairs, aims to improve coordination between local governments, urban search-and-rescue teams, and other international responders following a disaster. This tool basically provides real-time information sharing across the globe in the early phases of a disaster (Bjerge et al. 2016).

Sensor networks have been used widely for monitoring and surveillance applications. The advance of sensing technology has enabled the development of a low-cost, compact, and energy-efficient sensing device. Integrating it with processing and communication modules and linking them in a networking fashion have even augmented their capability in data-mining tasks, which gives huge incentives for more complex monitoring applications and a larger deployment area. In particular, wireless sensor networks have been proposed for disaster early warning, mitigation, and recovery due to their capability and flexibility to sense environment changes. For instance, Cayirci and Coplu (2007) proposed a sensor network for disaster relief operation management in Turkey, particularly due to earthquakes. Fantacci and colleagues (2010) integrated heterogenous network systems, including sensors and social networks, by using wireless communication infrastructure to support the public safety service in a smart city infrastructure. This work was conducted under a research project called In.Sy.Eme, for integrated system for emergency. This project combines two principal parts, efficient communication infrastructure to support information sharing and next generation grid principles (Fantacci et. al. 2009).

In disaster contexts, where timing and accuracy are critical and are combined with a harsh environment, limited power, and unreliable links, sensor network deployment becomes challenging and thus has been an attractive research topic for years. The focus of research may include, but is not limited to, optimal topology and routing, innovative data collection, and processing, minimizing energy consumption. Bahrepour and colleagues (2010) proposed a wireless sensor network to perform distributed event detection in the context of a disaster early warning application. They applied a machine learning technique that uses decision trees in a distributed manner to achieve a timely consensus derived from multiple detection events. da Silva and colleagues (2010) presented an energy-efficient mechanism for processing spatial queries on wireless sensor networks to detect danger in disaster situations. The proposed mechanism treats a region as irregular instead of rectangular shape when manipulating queries. The network should be independent, easily deployed, and adapted to different situations.

The remainder of this chapter is organized as follows. The next section elaborates the design and development of two components that constitute the system, namely a network of SBCs and D2D communication that uses mobile phones as the communicating devices. This section presents the requirements needed for the mobile device application to comply with the network's functionality. It also presents the alert dissemination scheme that is designed to work with the proposed network. Next is a description of the case study conducted to assess the network's performance. Specifically, we conducted a simulation of disseminating alert messages using the proposed network to investigate the timeliness of alert message delivery. It uses the AnyLogic simulation framework (AnyLogic Company 2018) to model alert message dissemination using the proposed system. We end by explaining the infrastructure contribution with respect to the current Indonesian early warning system.

Land-Based Network

In a larger perspective, our land-based network infrastructure consists of three main building blocks to support its operation: a web-based service called a community resilience framework (see chapter 1), a network of SBC devices, and a mobile application (see chapter 3). Figure 4-1 illustrates the general architecture of the network. This chapter focuses the discussion on the SBC network and D2D communication. It occasionally highlights certain issues in mobile applications that will emphasize the synergy between the network of SBCs and mobile phones.

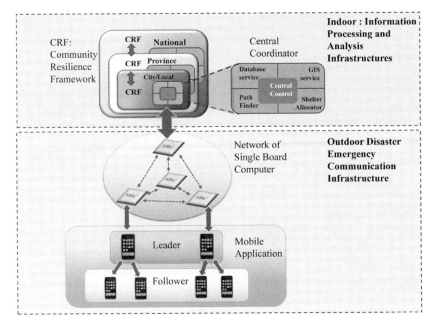

FIGURE 4-1. The architecture of the neighborhood information infrastructure.

Network of single-board computers

The network of SBCs provides minimum but sufficient communication access to the community at risk during disaster emergence. An SBC is a lightweight, inexpensive device, albeit a less sophisticated machine featured with a credit-card-size main board. Although the device has a small form due to advances in integrated circuity, it can perform computational tasks that a home desktop computer typically does at reasonable speed. Because the deployment of the network is mainly outdoors, having a small form offers an attractive feature that is needed to build a scalable, disaster-tolerant information infrastructure.

In this context of application, our network of SBCs is positioned as an emergency communication infrastructure and is not considered an on-par substitution to existing advanced communication infrastructures, such as cellular networks. Considering the Pi's capability, we designed the network to carry out small data traffic with decent latency. We tried to avoid large file transfers in the mobile application that runs on top of this infrastructure by making them available to the mobile phone during installation.

Building the prototype: SBC device. This project uses the SBC device called Raspberry Pi. Raspberry Pi was developed by the Raspberry Pi Foundation, a United Kingdom–based charity, to promote computing and put digital devices into the hands of people at affordable cost (Raspberry Pi Foundation n.d.). In November 2016, this device topped 11 million units sold (MagPi 2017). In 2017, Raspberry Pi sales had reached 12.5 million units, making it the third best-selling general-purpose computer (Miller 2017).

Raspberry Pi had released several models since February 2012. For our network prototype, we use Raspberry Pi 3 model B as the main component that constitutes the network. This model is powered by a 1.2 GHz, 64-bit quadcore ARM CPU, with 1 gigabyte memory. Compared to a conventional desktop, Raspberry Pi (Pi for short notation) provides more communication peripherals, such as a Universal Asynchronous Receiver Transmitter (UART), a Serial Peripheral Interface (SPI), USB, Bluetooth, and a general-purpose input-output (GPIO). The device is also equipped with an onboard IEEE 802.11.b/g/n/ac wireless LAN (commonly known as Wi-Fi) and an RJ-45 connector for Ethernet. Having multiple options for communication peripherals offers significant benefit and flexibility to develop input-output–extensive applications.

For this study, we built the network prototype that exploits IEEE 802.11 as the main technology to link among Pis. Because a Raspberry Pi is only equipped with a single onboard Wi-Fi adapter, three extra wireless network adapters are added to the device by making use of available USB ports.

Deployment. The SBC network uses static topology, and its deployment would mainly be outdoors. Pi would be placed in designated spots and needed to be planned accordingly. To give better support to community residents during evacuation, those spots may include roadsides along the evacuation path. Furthermore, because the network is expected to operate in disaster events, the Pi should be attached to a platform that is more resistant to damage due to disaster, such as an electric pole.

More work is needed to make the Pi ready to deploy, however. First, the Pi is not self-powered and thus should be accompanied with an external power source when being installed outside. Furthermore, placing the Pi on an electric pole could also give another advantage: in peacetime, the Pi may be powered by the electric pole itself as the primary power source. When disaster strikes and downs the main power source, backup power can take over to maintain the Pi operation. Because the Pi is only the size of a credit card, a commercially available power bank rated at 10,000 mAH is enough to keep the Pi working for roughly two to three hours for intense use with an onboard Wi-Fi adapter fully active. Indeed, larger backup power is needed if more powerful communication peripherals are used. Second, the Pi will be subject to weather, and an electronic device like that is prone to damage due to water ingression. Thus, the Pi and its accompanying components must be solidly protected within a weatherproof housing.

To cover a region, depending on the size of the region, one or more clusters of Raspberry Pi networks may be required. Each cluster has a dedicated Pi to communicate with the central coordinator (CC) who resides in the community resilience framework component. Communication to the CC can be made through satellite connection.

Finally, each Pi will be assigned an ID by the region to which it is belongs, called a Region ID. This ID is particularly important whenever an alert needs to be disseminated to a certain region. Alert messages will be routed to the Pi whose Region ID matches the one incorporated in the alert message.

Configuration and firmware development. A Raspberry Pi has a fair number of options for its operating system. We choose Raspbian as the operating system (OS), as it is free and is developed based on Debian Linux. This version of OS is fairly popular among Raspberry Pi developers. Various applications and libraries are available from an external repository under this OS. It also provides more compatibility with various peripherals in practice.

For our application, we developed firmware that runs in the background in each Raspberry Pi to perform the infrastructure's core functions. The firmware architecture is designed and developed using a multithread, producer-consumer paradigm. Because the Raspberry Pi's kernel is Linux based, which was initially developed using C language, we selected C language programming to develop the firmware.

With a multithread design principle, we dedicate each thread to perform a specific function. Internal communication and coordination among modules in the firmware is performed through a shared queue. This design principle will allow easier modification when a new functionality needs to be accommodated in the future.

The firmware consists of three modules, namely two Communication modules and one Function module. The Communication module performs data packet reception and transmission over the link. The Communication module should be developed based on link technology that will be selected for this task. The technology may include, but is not limited to, RF link, Wi-Max, IEEE 802.11, and ZigBee. In our network prototype, IEEE 802.11 technology standard is used at the link layer, combined with an IP-based socket at transport layer. The Function module comprises multiple worker submodules with their respective first in–first out (FIFO) queues, which are also shared with the Communication module. The Worker submodules along with their associated buffers are differentiated based on incoming data type.

As the packet arrives at the Pi, the Communication module at the reception side reads the packet and puts it into its respective Worker's FIFO queue in the Function module. The designated worker picks the data from the queue and processes it accordingly, assembles a new packet if necessary, and puts it into the Communication module's outgoing side. The interactions between those modules resemble the producer-consumer principle. The challenge in making this principle work properly is to guarantee atomic access of the queue. Only one module can gain access to the queue at a time, before the other takes a turn. Therefore, to ensure atomic access, mutually exclusive and semaphore variables are used. These variables can only be used by one process until it releases it. The other challenge is setting the queue size. The queue size should be large enough to accommodate the burst of incoming traffic in the worst case.

With this architecture in place, Raspberry Pi firmware manages two main tasks: application-related tasks and mobility management tasks. These two tasks are developed tightly in accordance to the firmware implemented in mobile phones. In the application-related task, Raspberry Pi mainly does packet forwarding and routing. If necessary, Raspberry Pi may make minor modifications to the content of the packet before forwarding it to its ultimate destination. Concerning the mobility management task, the firmware runs a connection and tracking scheme to ensure that the packet is delivered to the desired mobile phone while the mobile phone moves.

Maintaining mobile communication in the network of SBCs poses formidable challenges despite the network's small scale and limited capability to carry the traffic. As for comparison, in cellular networks, the phone's location is monitored and managed continuously by base stations, mobile switching centers, and home location registers. Mobility in management in such a massive system is backed up by powerful database machines and backboned with a high-speed link, which is not the case in this infrastructure. SBCs have limited computational power, even compared to traditional PCs.

Therefore, our proposed network does not rely on a single centralized user's location database server. We do not dedicate a single Raspberry Pi to be the location database server in a region. Instead, we store user locations in several Pis that act as database servers. To perform a database service in each Pi while mindful of the Pi's limited computational capacity, we installed a lightweight database engine called SQLite.

Mobile phone location tracking scheme

As mentioned previously, our proposed network implements a distributed location database server due to limitations of the Raspberry Pi's computational and storage capacity. In essence, a mobile phone will be assigned a permanent location database Pi called the *Home Agent Pi.* The *Home Agent Pi* is responsible for keeping track of the Raspberry Pi's ID in which the mobile phone is currently located. Because a non-single-location server might be present in a region, different mobile phones in the same region might have different *Home Agent Pis.* There are two additional record fields maintained during mobile phone tracking, *Current Pi* and *Previous Pi.* As the terms infer, the field *Previous Pi* is the ID of the Raspberry Pi from which the mobile phones come, whereas *Current Pi* represents mobile phone's current location.

In addition to maintaining mobile phone location information, the tracking scheme also caches packets at the destination Pi. Upon packet arrival at the destination Pi before being delivered to the mobile phone, the packet would be cached to anticipate if the intended mobile phone is not within coverage. The packet would be retrieved from the cache upon receipt of the release cached packet command sent by another Pi. The lifetime of the packet within the cache is not unlimited and is determined by a timer. When the timer expires, the content of the cache will be flushed. Setting the duration of the timer should consider the human speed of moving and the average distance between Pis in a region. A denser Pi network results in shorter mobile phone transit time from one Pi to another and hence requires a shorter timer duration.

The mobile phone actively updates its location at regular intervals and reports its location to its *Home Agent Pi* as necessary. When a mobile phone enters a Pi's coverage, the mobile phone queries the Pi to obtain the ID of the Pi. This Pi is referred to as *Current Pi*. Once the mobile phone gets the *Current Pi*'s ID, the mobile phone reports its current location to its *Home Agent Pi*. Upon receiving the report, the Home Agent updates the mobile phone's entry in its database. Subsequently, the *Home Agent Pi* sends an acknowledgment packet to the *Current Pi*, containing the information about the ID of the *Previous Pi*. Using this information, the *Current Pi* will send a request to the *Previous Pi* to release any cached packets for the corresponding mobile phone. Figure 4-2 illustrates the tracking scheme implemented in this proposed network.

Importantly, the Pi does not maintain the connection state of every mobile phone. A logical connection between mobile phones, if needed, must be established at the application layer. This design makes our tracking scheme much simpler in practice. Furthermore, our tracking scheme does not guarantee reliable delivery. It means that reliable data transmission must be made at the application layer as well.

Link, network, and transport protocol

Each Pi has one onboard 802.11 b/g/n/ac (Wi-Fi) dual band wireless LAN adapter. To create interconnections between Pi's in our prototype, we need to rely on more Wi-Fi ports. Therefore, we use available USB ports to provide more Wi-Fi ports, using a USB to 802.11 adapter. Including onboard Wi-Fi, we configure four Wi-Fi ports for this purpose. Three of them are used to connect the Pi with the others, while the fourth port is configured to provide the hotspot needed by the mobile

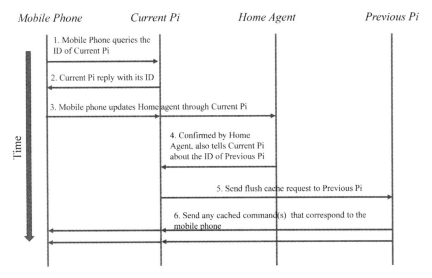

FIGURE 4-2. Tracking scheme in our proposed network.

phone to access the network. In some cases, depending on the network topology, we may have one of those three ports unused for interconnection, so we can reserve it for monitoring and development purposes. We assign a name for each wireless adapter— *wlan0, wlan1, wlan2,* and *RaspIF*—for identification purposes.

We use an Edimax EW-7811Un 802.11 b/g/n Wi-Fi nano USB adapter to configure the available USB ports to be a Wi-Fi wireless Ethernet. This device is one of only a few adapters that can be configured as either an access point or a wireless Ethernet. It is also supported by the Pi's internal driver under Raspbian OS. Controlled through a utility called hostapd, for host access point daemon, we created a Wi-Fi hotspot using this device. Hostapd is a user space software access point capable of turning normal network interface cards into access points and authentication servers (http://w1.fi/hostapd/). We dedicate *wlan1* port to provide a Wi-Fi hotspot to mobile phones, and the hotspot has the same service set identifier designation (SSID) in each Raspberry Pi. In essence, SSID is the literal name for the Wi-Fi hotspot.

To create a physical link with other Pi, we configured *wlan2* as an active adapter, also by using *hostapd* utility, while *wlan0* or *RaspIF* are configured as passive ports. Because *wlan2* is configured as an active port, it provides a Wi-Fi hotspot used by another Pi to connect through its adapters using either *wlan0* or *RaspIF*. Having another active wireless adapter that can be configured as additional access point also allows us

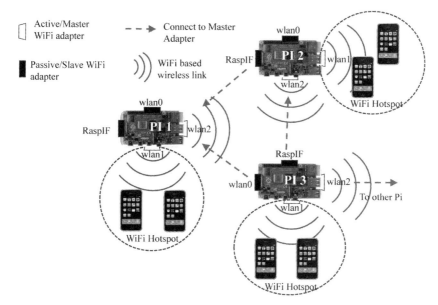

Figure 4-3. Configuration of Pi network.

to create branch in our network topology. Figure 4-3 shows an example of how to arrange the Pi interconnection using those wireless adapters.

In addition to SSID, an active Wi-Fi port is also identified through its basic service set ID (BSSID). A BSSID is the hardware address of the wireless interface and is unique from one device to another. Normally, a BSSID cannot be modified, whereas an SSID is a user-defined identity. Raspbian OS, however, provides a way to manipulate a BSSID. With this benefit, we investigated three different Wi-Fi hotspot configurations as follows: (1) each Raspberry Pi has a different SSID and BSSID, (2) each Raspberry Pi uses the same SSID but a different BSSID, and (3) each Raspberry Pi uses the same SSID and BSSID. We found that scheme 2 gives a slightly better association time than scheme 1 because the mobile phone does not need to check its internal database. Scheme 3 actually gives the fastest association time. The connection between the mobile phone and Raspberry Pi becomes inconsistent if the Pi coverage overlaps one another, however, because the mobile phone cannot distinguish the correct Pi to which it associates. Therefore, we selected scheme 2 for our proposed network.

At the network layer, typically private internet protocol (IP) addresses will be used for a local hotspot. We made a different approach for the IP address assignment for mobile phones and the Pi within Pi hotspot controlled through the interface *wlan1*, however, and instead of using a private IP address, we use IPv4 class D address space for communication purposes. This address space is reserved for multicast communication. In this scheme, each smartphone and Pi will be assigned the same multicast IP address group (Cotton et al. 2010). This scheme enables a broadcast-like communication, but it is restricted to mobile phones and Raspberry Pi with the same multicast IP address.

We choose an IP multicast addressing scheme due to following reasons. First, having the same static multicast IP address across mobile phones avoids the use of dynamic IP assignment through the dynamic host configuration protocol (DHCP) (Droms 1997) service. Although DHCP gives an advantage in terms of flexibility on IP address assignment, having the DHCP service running within the hotspot may slow down the mobile phone association process to the Pi. Because each mobile phone has the same multicast IP address, a unique ID must be determined at the application level to identify the mobile phone. In this case, managing mobile phone IDs at the application level is easier than at the transport layer. Second, during our development, we occasionally found that some mobile phones do not support the broadcast IP address (255.255.255.255) anymore, and sometimes the links are unreliable. The mobile phone sometimes cannot receive the data broadcasted with this broadcast IP address even if the data were transmitted in near proximity. In contrast, we found that communication is more stable when using the multicast IP address.

Different from *wlan1*, we assigned a predetermined set of local IP addresses. We allocated local IP addresses to the remaining wireless interfaces (*wlan0*, *RaspIF*, and *wlan2*), making the network inherently inaccessible from public internet infrastructure. The only access to the network is through the Pi, which connects to the CC. We have prepared four sets of local IP addresses so that each set has a different subnet. In the Pi, each adapter does not share the same subnet with another. The link arrangement is made so that the IP address of the active and passive interfaces in a Pi-to-Pi link are always in the same subnet. Furthermore, we prepared different SSID and channel allocations on *wlan2* for each subnet. When we connect *wlan2* with other wireless interface (*RaspIF* or *wlan0*), we manage this link so that the subnet, channel, and SSID for this connection would be sufficiently distant from any other Pi-to-Pi interconnection that uses the same subnet, channel, and SSID. The separation is interleaved at least by three Pi units. Although this arrangement may not be highly critical in actual deployment, we designed this link arrangement for our network prototype for the following reasons. First, such a link arrangement makes managing, observing, and testing Pi to Pi interconnections easier, especially during the development phase. Second, we avoid interference caused by nearby Pis, especially during the development phase. We sometimes need to put multiple Pis in near proximity to each other, which can result in coverage overlap. If this link (*wlan2-RaspIF/wlan0*) shares the same channel, subnet, and SSID, it becomes difficult to manage the topology and monitor the traffic.

At transport layer, we utilize user datagram protocol (UDP) transport protocol. The advantage of UDP is that it is simple and does not require connection establishment. This feature gives a significant advantage, especially when connect-disconnect events occur frequently as the mobile phone moves across different Pi coverages. Transmission can take place immediately after the mobile phone physically associates to a Pi's Hotspot through *wlan1* interface. Furthermore, this transport protocol is more stable, using a multicast addressing scheme. It is not reliable, however, because no retransmission is requested in the case of failure packet delivery (due to packet error or loss). Consequently, we define our own retransmission scheme at the application level so as to increase link reliability.

When the link between Pis is established and up, each Pi runs a small utility to build its routing table. This utility basically runs a distance vector routing protocol using the Bell-Ford algorithm (Bellman 1958) to build the table. This utility also built a broadcast table in each Pi. This table contains an adjacent parent and child Pi as it is seen from the Pi. Collection of the table data forms a snapshot of the overall structure of the tree topology as it is created from the cluster-head Pi, the Pi which is linked to the CC. As the result, an adjacent neighbor Pi does not necessarily imply a parent or child Pi from a Pi. This table is particularly useful to help broadcast the information across multiple Pi within the network efficiently, as it can prevent unnecessary broadcasts and hence mitigate broadcast storm. If a Pi receives a broadcast-type packet (an alert message), it only forwards the packet to its parents or child.

We do not extend the scheme to the periodic routing table update. For this prototype, we assume that no link failure occurs between Raspberry Pis.

Information delivery

The land-based network carries out two types of information delivery schemes, one to one and one to many. The former type carries the information between mobile phones and is designed to establish one-to-one communication between mobile phones. The latter type is designed to support dissemination of alerts to neighborhood leaders from the emergency operations center that manages the community resilience framework (see figure 4-1).

One to one. In one-to-one type of delivery, most of the information carried within the network is from mobile phone to mobile phone, such as emergency messaging during a disaster event. Because mobile phones move across different Pi, mobile phone location tracking plays a crucial role to carry out this delivery properly. When a mobile phone (sender) wants to reach another mobile phone (destination), it needs to contact the *Home Agents* of the destination mobile phone (shown at step 1 of figure 4-4). The *Home Agent* will inform the sender about the current location of the destination (step 2 of figure 4-4). The sender uses this information to send the message to the destination (step 3 of figure 4-4).

If the destination needs to reply, either of two approaches can be applied. The first approach requires the destination to reply the same way as the sender did, by contacting the sender's *Home Agent*, whereas the second approach imposes the sender to incorporate its current location into the message that will be exchanged. Steps 3 and 4 of figure 4-4 illustrate the second approach. Without loss of generality, figure 4-4 assumes that the sender is mobile. When the sender moves to Pi 2 and after it updates its *Home Agent*, it can include its most recent location to the message, as shown in step 5 of figure 4-4. This approach is better in some situations, particularly when the sender and the destination need to maintain their message exchanges within a limited time. Incorporating recent location information into the message also implies that the location inquiry step to the *Home Agent* be skipped, hence reducing the *Home Agent*'s load on processing location inquiries.

One to many. The one-to-many type of delivery is mainly developed to distribute alert messages to a community at risk. This emergency infrastructure uses two schemes to inform the community: multicast in Pi network and D2D communication.

In our proposed application, an alert will be sent only to the area that will be severely impacted by the disaster. To conduct such mechanism, the alert message will be tagged with the ID of the high-risk region. When an alert message is issued to a certain region ID, the message will be disseminated through the Pi with the same region ID.

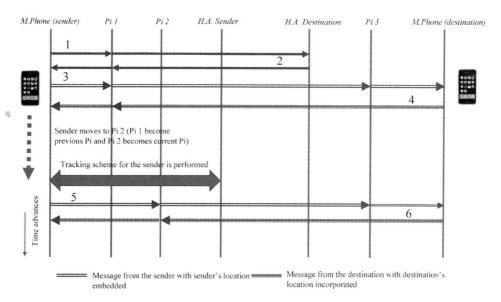

FIGURE 4-4. One-to-one message delivery mechanism.

Because this infrastructure is designed to adapt with community so-cial structure, an alert message will be sent to the leader of that region. The leader then initiates an alert dissemination to followers. We define two steps for alert dissemination in this context. In the first step, the alert message from the leader is disseminated through the Pi network. In this step, the broadcast table described above comes into play to disseminate the alert.

Every community resident who had gotten the alert message becomes an agent for alert distribution. At this moment, the second step formally commences, and a D2D communication scheme disseminates the alert message. Technically speaking, information broadcasted by mobile de-vices will be subsequently repeated by other mobile devices that receive the message. This mechanism is particularly useful to extend alert de-livery to mobile phones that are not currently within a Pi coverage, but it could potentially result in broadcast storm and frequent message col-lisions. Therefore, we carefully design how the alert message should be distributed to affected community residents by assigning each mobile phone to pick a random timeout value before initiating transmission. In our scheme, we set a range of timeout values according to the Pi's hotspot radius and people's movement speed. Figure 4-5 illustrates the workflow of the scheme in more detail.

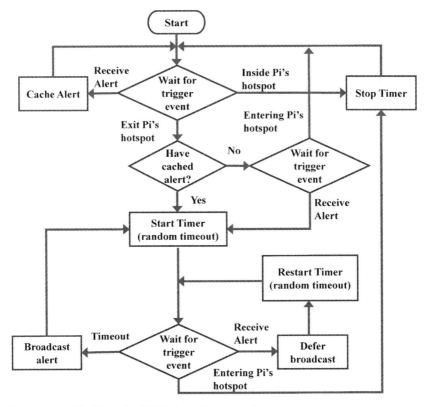

FIGURE 4-5. Workflow for D2D alert dissemination.

Each mobile phone regularly checks its presence status within a Pi's coverage. When the mobile phone is inside the Pi's hotspot, it only listens and caches incoming alert messages. When the mobile phone steps out of the Pi's hotspot and it has cached alert messages, it starts its random timer while it keeps listening to anticipate new alert messages. If it overhears an alert message while it is outside of the Pi hotspot, it defers transmission and restarts the timer again. If it does not overhear an alert message and the timeout expires, it will broadcast the message to any neighboring mobile phones within its radio range.

Modeling for Information Dissemination

In this section, we develop a simulation scenario to assess the performance and feasibility of the proposed SBC network to deliver the alert message. The goal of the simulation is to investigate the timeliness aspect of alert delivery given a certain number of mobile phones and Raspberry Pi within an area. In this simulation, we apply the scheme explained above using an AnyLogic agent–based simulator to model alert message dissemination to community residents. We built the simulation scenario using Java language.

Defining agents

The important step in developing this simulation is to define agents properly. In AnyLogic, an agent can be seen as an object or actors that perform a certain task. The agent can also be interpreted as an object that holds a collection of attributes. AnyLogic provides a reasonable number of tools and methods that are needed to model interaction between agents. Communication between agents is performed through message passing methods. The following sections describe briefly the agents that are defined for our simulation.

Main. Whatever the simulation will be, an object called Main is created by default in AnyLogic. This agent is responsible for holding events, environments, and actors that are defined in a simulation. In the Main agent, other user-defined agents, events, and environments are initialized. Because the Main agent encapsulates entire user-defined agents, the model dynamics aspects such as simulation statistics, progress over time, and changes of attributes can be observed and collected from this agent.

Raspberry. The Raspberry agent represents Raspberry Pi devices. During simulation, we created multiple instances of Raspberry Pi to be deployed within the environment in the Main agent. We define this agent to behave in the same way as our actual Raspberry Pi device. We also define a method to allow two-way communication between Raspberry Pis, Raspberry Pi, and Human agent (to be explained subsequently).

Human. The Human agent represents mobile phone objects. We created multiple instances of the Human agent in the Main agent environment and spread their locations randomly within the environment. To model human mobility, the agent moves randomly in direction and distance, using a preset walking speed as the simulation begins. We repeat random movement assignment each time the agent completes the move. While moving, the agent keeps listening for messages and is only reachable when they are within radio range. We define two subtypes of Human agent in our simulation: leader and follower. These attributes are provided to reflect the social structure of the community in our case study.

Channel. As the name implies, the Channel agent represents wireless channel. Because AnyLogic does not provide a wireless channel model in its library, we must define our own wireless channel agent. When the Human or Raspberry agent begins to communicate, it needs to check the Channel agent to see if the outcome of the link fails or succeeds. We also define a simple medium-access scheme that mimics carrier sense multiple access. This medium-access scheme is widely used in wireless communication. In simple words, this scheme requires an agent to check the channel whether the channel is occupied or not before sending the message. If the channel is occupied, the agent needs to wait for another random timeout before it starts checking the medium again. We also make the scope for medium sensing to be radio range–dependent, which means that any two distant agents can transmit simultaneously if their interdistance exceeds their radio range.

Frame. The Frame agent represents the packet format that will be exchanged during simulation. This agent is essentially a data structure that comprises several fields. Some of the core fields defined in this agent are *Command*, *SourceID*, *DestinationID*, *PackageID*, *Options*, *Payload*, and *Timestamp*. The *Command* field denotes the content type that the message carries. The *SourceID* and *DestinationID* fields denote the ID or the source and destination, respectively. The *PackageID* field is used to identify the packet. Together with the *Timestamp* field, the *PackageID* is important for packet tracking, especially when we investigate the timing of the packet during a performance test. The *Options* field is required to store additional information about the packet and is particularly needed when a packet content is modified during transit, whereas the *Payload* field carries the main content of the packet.

Defining simulation scenario and parameters

The simulation uses a predefined area with size of 500 meters by 500 meters, or 0.25 square kilometer. We place Raspberry Pi agents in a grid topology with Pi interdistance as the parameter. We vary the distance between Pi to model different Pi networks' density in a region. We also vary the density of the Human agent within the predefined area to investigate the effectiveness of our alert dissemination scheme. According to Badan Pusat Statistik Kota Padang (2018), Padang City's population density is about 1,370 people per square kilometer. It is equivalent to 350 Human agents in our simulation area. For this simulation, we vary the number of followers between 100 to 350 people.

We use normal distribution for both Human agent distribution and Human agent movement pattern. According to Bumgarder (2022), human fast-walking speed is about 1.79 meters per second. For our simulation, we round it up to 2 meters per second. The reason we selected fast-walking speed is because the message dissemination will occur in a disaster emergency situation, where we assume people will have a tendency to walk faster.

Our internal test demonstrated that our tracking scheme implemented in Raspberry Pi needs about 10 milliseconds on average to process the mobile phone's location. Without loss of generality, we also assume that this duration is a fairly reasonable processing time for other tasks performed by Raspberry Pi. Our internal test also demonstrated that the Raspberry Pi wireless range outdoors varies between 30 to 50 meters. The test shows that at 30 meters, we gained more than 90 percent of successful transmissions on average (see chapter 7). With these test data, we decided to set 40 meters as the maximum radio range in the simulation. This radio range is used to characterize our Raspberry Pi and mobile phone coverage. Finally, we ignore radio propagation time because the time it takes to travel within our environment would be negligible.

We also use random timeout in D2D communication and compute it based on human walking speed and mobile phone radio ranges. With 40 meters of radio range, we computed the timeout value to be normally distributed within an interval between 5 to 15 seconds.

Results and analysis

The goal of this simulation is to investigate how fast an alert disseminated, given the varying number of Raspberry Pi and population density in an area. Two scenarios have been prepared. In the first scenario, we selected five different Pi interdistance values: 500, 400, 200, 125, and 100 meters. In grid-topology arrangement, they are roughly equal to the density of 1, 4, 9, 16, and 25 Raspberry Pis, respectively, in 0.25 square kilometer. Our simulation starts with one leader who has 349 followers in our simulation environment.

Figure 4-6 shows the average number of people who received the

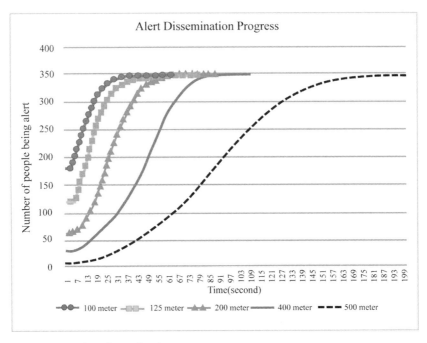

FIGURE 4-6. Alert dissemination progress.

alert message over some interval of time given a particular number of Raspberry Pis. The figure essentially demonstrates that having more Pis can reduce alert dissemination time. The result reveals an interesting characteristic, however. The result exhibits an S-shaped pattern. Our simulations exhibit the flattening curve after at least more than three hundred people received the alert.

Columns 2 and 3 of table 4-1 show the average time and standard deviation to alert the community completely. We observed that about five unalerted people remained within this deviation, which constitutes about 1.5 percent of the community. Alerting the remaining people takes about 10 to 25 percent of the total dissemination time to complete. For this reason, we measure the dissemination time required to confirm that at least 95 percent of the community residents have received the message. We count this number as the worst-case assumption. In our simulation, this conservative, worse-case assumption leaves about seven people remaining unalerted, which we consider reasonably acceptable in practice. As shown in table 4-1, the time it takes to accomplish dissemination that reaches at least 95 percent of the community residents falls between 40 to 75 percent of the total time needed to alert 100 percent of the residents.

TABLE 4-1. Dissemination time average and standard deviation

Pi Interdistance (meters)	100 Percent Alerted		95 Percent Alerted	
	Average Time (seconds)	Standard Deviation	Average Time (seconds)	Standard Deviation
500	201.9	21.7	153.1	12.6
400	107.0	17.6	72.3	5.7
200	89.1	17.4	46.9	3.7
125	78.5	16.8	37.0	4.9
100	69.6	15.3	28.5	3.7

Increasing the number of Pi devices indeed improves the alert dissemination time, especially the time to reach more people at the first step of the alert distribution scheme as described below. The time reduction is less significant as the number of Pi increases. For instance, reducing the Pi interdistance from 500 to 400 meters, which implies four times the increase in terms of the number of Raspberry Pi, only reduces the time to about a half of total dissemination time at 95 percent alerted. With this result, we can see the potential use of our Raspberry Pi network for emergency communication infrastructure, especially to disseminate alert messages.

These results are obtained with the assumption that the leader is always getting the alert in the first place. If that is not the case, the dissemination time could take longer. Figure 4-7 depicts one example of the dissemination progress when the leader is not reachable at the first attempt. The figure demonstrates some delay before the alert signal starts to disseminate rapidly. To overcome this problem, more leaders are indeed required to ensure that at least one leader in each particular region received the alert from an emergency operation center. If necessary, the leader selection scheme as described in chapter 3 may be performed. Another approach that can be done is to make the Pi repeat broadcasting the alert at regular intervals.

In the second scenario, we vary the density of people to see how effective the proposed D2D communication scheme is. We fixed the number of Raspberry Pi devices at nine units per 0.25 square kilometer and varied the number of the people from 100 to 350 in increments of 50 people. Figure 4-8 depicts the total dissemination time for 95 percent of the community residents having received the alert message.

As we expected, the effectiveness of this scheme decreases as we have less dense people. Duration, however, does not increase significantly

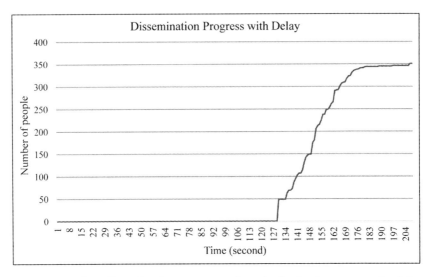

FIGURE 4-7. Alert dissemination progress when the leader has not received the first attempt.

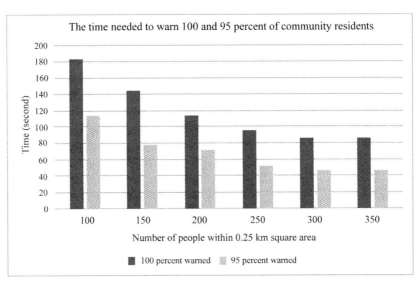

FIGURE 4-8. The time needed to have 100 percent and 95 percent of people warned.

between 250 to 350 people, given nine units of Pi within the simulation area. The dissemination time starts increasing rapidly when the scheme is applied in an area with density of fewer than 200 people. To overcome this problem, we may need to add Raspberry Pi devices to reach more people.

Integration with the Indonesian Current Tsunami Early Warning System

To demonstrate the importance of our proposed system to assist the flow of critical information during disaster, we use Indonesia as case study. Indonesia, as the largest archipelago country, has intensified its efforts to mitigate the impact of natural hazards. It has the world's fourth largest population, with diverse ethnic groups and varied social cultures and economic backgrounds, and the size and scale of the country have increased the complexity of the effort.

The issue of disaster preparedness has gained serious attention since the megatsunami disaster hit Indonesia in 2004, causing a death toll of more than one hundred thousand people. Indonesia is geographically located in a zone called the "Ring of Fire" and subduction zone, making it prone to earthquakes. If an earthquake occurs in the sea, it may trigger a tsunami.

To anticipate and mitigate the impact of tsunamis, Indonesia launched a national system that encompasses the development and deployment of seismographs, tide gauges, buoys, and information dissemination infrastructure that are integrated into a national disaster operations system known as the Indonesia Tsunami Early Warning System (InaTEWS). See chapter 2, figure 2-2, for the architecture and information flow of InaTEWS.

In the context of InaTEWS, our proposed approach does not offer a complete substitute for the existing system. Instead, our infrastructure can provide alternative communication means to communities at risk. Based on figure 2-2, the application of this information infrastructure is relevant at the downstream level. This proposed infrastructure provides a means to disseminate the alert that is triggered by the decision support system (DSS) unit that is more adapted to the affected community. Our network SBCs, coupled with D2D communication, could provide the means to conduct notifications and warning downstream.

With the InaTEWS system already in place, integrating our proposed network with it should be managed and assessed accordingly. We see that the potential point for coupling these two different infrastructures lies within the data and information analysis layer, which likely fits best the responsibilities of the national center. Because this component is part of the land-based network and is concerned with analyzing and visualizing the dynamic events during disaster emergence, we developed a web-based community resilience framework. Further detailed description of this framework is found in chapter 1.

We expect the outcomes of this research to offer the following benefits in conjunction with InaTEWS:

- The proposed system is not a substitute for the currently existing early warning system. Instead, it would complement the existing system, which will improve the evacuation process during tsunami events.
- This design will augment the existing tsunami early warning system and provide options to local governments to extend the warning system while keeping its cost within reasonable economic limits.
- The most important contribution of this infrastructure for the current InaTEWS data flow is to provide a feedback mechanism from the downstream level to higher levels, especially to the DSS unit. Real-time information sharing between a community and the DSS unit is essential to make more informed decisions and resource allocations.

Conclusion

Building community resilience to hazards has shifted to a new paradigm that engages the whole community in active participation by ensuring real-time information exchange and knowledge sharing between community residents and government agencies legally responsible for disaster mitigation and response operations. Developing timely, informed collaboration between local social networks and disaster operations personnel represents a continuing need for communities at risk. To meet this need, we proposed a scalable, disaster-tolerant, socially aware infrastructure that combines a network of SBCs with mobile applications and Java-based controller.

Our Raspberry Pi network combined with a D2D communication scheme shows the potential application for disaster emergency communications. With the current population density of Padang City, the dissemination time is under 3 minutes at the longest, which is still acceptable to alert at least 95 percent of community residents in practice (see simulation and test results presented in chapters 6 and 7). Large-scale network deployment is still needed to verify the actual performance of the proposed network in practice.

In the context of the current Indonesian early warning system, this infrastructure can provide more options for communication to the downstream level. Further, the infrastructure could also provide feedback mechanisms to DSS units to support a more informed, timely decision-making process during a disaster.

References

Ai, Fuli, Louise K. Comfort, Yongqiang Dong, and Taieb Znati. 2016. "A dynamic decision support system based on geographical information and mobile social networks: A model for tsunami risk mitigation in Padang, Indonesia." *Safety Science* 90: 62–74.

Ali, Kamran, Huan X. Nguyen, Quoc-Tuan Vien, and Purav Shah. 2015. "Disaster management communication networks: Challenges and architecture design." In *2015 IEEE International Conference on Pervasive Computing and Communication Workshops (PerCom Workshops)*, 537–42. IEEE.

AnyLogic Company. 2018. "Simulation modeling software tools and solutions for business." Retrieved November 2018 from www.anylogic.com.

APCICT. 2011. *Academy of ICT Essentials for Government Leaders: Module 9: ICT for Disaster Risk Management.* United Nations Asian and Pacific Training Centre for Information and Communication Technology for Development, UN-APCICT/ESCAP. https://www.unapcict.org/sites/default/files/2018-12/BN%209.pdf.

Arneson, Erin, Derya Deniz, Amy Javernick-Will, Abbie Liel, and Shideh Dashti. 2017. "Information deficits and community disaster resilience." *Natural Hazards Review* 18, no. 4: 04017010.

Badan Pusat Statistik Kota Padang (Padang City Central Bureau for Statistic). 2018. "Statistik Daerah Kota Padang Tahun 2018." Catalogue: 1101002.137, ISBN: 978-602-0712-10-9. https://padangkota.bps.go.id/publication/2018/10/01/0e565b67e0130e18ee037cf1/statistik-daerah-kota-padang-2018.html.

Bahrepour, Majid, Nirvana Meratnia, Mannes Poel, Zahra Taghikhaki, and Paul J. M. Havinga. 2010. "Distributed event detection in wireless sensor networks for disaster management." In *2010 International Conference on Intelligent Networking and Collaborative Systems*, 507–12. IEEE.

Bellman, R. 1958. "On a routing problem." *Quarterly of Applied Mathematics* 16, no. 1: 87–90.

Bjerge, Benedikte, Nathan Clark, Peter Fisker, and Emmanuel Raju. 2016. "Technology and information sharing in disaster relief." *PloS One* 11, no. 9: e0161783.

Bruneau, Michel, Stephanie E. Chang, Ronald T. Eguchi, George C. Lee, Thomas D. O'Rourke, Andrei M. Reinhorn, Masanobu Shinozuka, Kathleen Tierney, William A. Wallace, and Detlof Von Winterfeldt. 2003. "A framework to quantitatively assess and enhance the seismic resilience of communities." *Earthquake Spectra* 19, no. 4: 733–52.

Bumgardner, Wendy. 2022. "What is a brisk walking pace? How to boost your average walking speed

for more exercise benefits." Verywell Fit. https://www.verywellfit.com/how-fast-is-brisk
-walking-3436887.

Cayirci, Erdal, and Tolga Coplu. 2007. "SENDROM: Sensor networks for disaster relief operations
management." *Wireless Networks* 13, no. 3: 409–23.

Comfort, Louise K. 2005. "Risk, security, and disaster management." *Annual Review of Political Science*
8: 335–56.

Comfort, Louise K. 2007. "Crisis management in hindsight: Cognition, communication, coordination, and control." *Public Administration Review* 67: 189–97.

Comfort, Louise K. 2019. *The Dynamics of Risk: Changing Technologies and Collective Action in Seismic
Events.* Princeton University Press.

Comfort, Louise K. 2020. "Managing critical infrastructures in crisis." In *Oxford Research Encyclopedia
of Politics.* https://doi.org/10.1093/acrefore/9780190228637.013.1646.

Comfort, Louise K., and Thomas W. Haase. 2006. "Communication, coherence, and collective
action: The impact of Hurricane Katrina on communications infrastructure." *Public Works
Management and Policy* 10, no. 4: 328–43.

Comfort, L., Daniel Mosse, and Taieb Znati. 2009. "Managing risk in real time: Integrating information technology into disaster risk reduction and response." *Commonwealth: A Journal of Political
Science* 15, no. 4: 27–45.

Comfort, Louise K., and Haibo Zhang. 2020. "Operational networks: Adaptation to extreme events
in China." *Risk analysis* 40, no. 5: 981–1000.

Cotton, Michelle, Leo Vegoda, and David Meyer. 2010. *IANA Guidelines for IPv4 Multicast Address
Assignments.* RFC 5771, March 2010. https://www.rfc-editor.org/rfc/rfc5771.txt.

da Silva, Rone Ilídio, Virgil Del Duca Almeida, Andre Marques Poersch, and Jose Marcos Silva
Nogueira. 2010. "Wireless sensor network for disaster management." In *2010 IEEE Network
Operations and Management Symposium-NOMS 2010,* 870–73. IEEE.

Droms, Ralph. 1997. "Dynamic host configuration protocol" (No. RFC 2131). Retrieved from
https://tools.ietf.org/html/rfc2131.

Fantacci, Romano, Dania Marabissi, and Daniele Tarchi. 2010. "A novel communication infrastructure for emergency management: The In. Sy. Eme. vision." *Wireless Communications and Mobile
Computing* 10, no. 12: 1672–81.

Fantacci, Romano, Marco Vanneschi, Carlo Bertolli, Gabriele Mencagli, and Daniele Tarchi. 2009.
"Next generation grids and wireless communication networks: Towards a novel integrated
approach." *Wireless Communications and Mobile Computing* 9, no. 4: 445–67.

Federal Emergency Management Agency. 2011. *A Whole Community Approach to Emergency Management:
Principles, Themes, and Pathways for Action.* Report Number: FDOC 104-008-1. Washington, DC:
Federal Emergency Management Agency. https://www.fema.gov/sites/default/files/2020-07
/whole_community_dec2011__2.pdf.

Federal Emergency Management Agency. 2014. *Effective Communication, Student Manual, Lesson 3: Communicating in an Emergency,* IS-242.b.

Fountain, Jane. 2001. *Building the Virtual State: Information Technology and Institutional Change.* Washington, DC: Brookings Institution Press.

Graber, Doris. 2002. *The Power of Communication: Managing Information in Public Organizations.* Washington, DC: CQ Press.

Kapucu, Naim. 2006. "Interagency communication networks during emergencies: Boundary spanners in multiagency coordination." *American Review of Public Administration* 36, no. 2: 207–225.

Klein, Gary A., Judith Ed Orasanu, Roberta Ed Calderwood, and Caroline E. Zsambok. 1993. *Decision Making in Action: Models and Methods.* Ablex.

MagPi. 2017. "11 Million Sold, Raspberry Pi Sales Pass Another Major Milestone." *MagPi* 53 (January): 10. Retrieved December 2018 from https://www.raspberrypi.org/magpi-issues/MagPi53
.pdf.

McCloskey, John, Andrea Antonioli, Alessio Piatanesi, Kerry Sieh, Sandy Steacy, Suleyman S. Nalbant, Massimo Cocco, Carlo Giunchi, Jian Dong Huang, and Paul Dunlop. 2007. "Near-field propagation of tsunamis from megathrust earthquakes." *Geophysical Research Letters* 34, no. 14.

Mileti, Dennis. 1999. *Disasters by Design: A Reassessment of Natural Hazards in the United States.* Joseph Henry Press.

Miller, Paul. 2017. "Raspberry Pi Sold Over 12.5 Million Boards in Five Years." *Verge*, March 17, 2017.   https://www.theverge.com/circuitbreaker/2017/3/17/14962170/raspberry-pi-sales -12-5-million-five-years-beats-commodore-64.

National Research Council. 2012. *Disaster Resilience: A National Imperative.* Washington, DC: National Academies Press. https://doi.org/10.17226/13457.

Nowell, Branda, and Toddi Steelman. 2015. "Communication under fire: The role of embeddedness in the emergence and efficacy of disaster response communication networks." *Journal of Public Administration Research and Theory* 25, no. 3: 929–52.

Raspberry Pi Foundation. n.d. "About Us." Raspberry Pi Foundation. https://www.raspberrypi.org /about/.

Sasorova, E. V., M. E. Korovin, V. E. Morozov, and P. V. Savochkin. 2008. "On the problem of local tsunamis and possibilities of their warning." *Oceanology* 48, no. 5: 634–45.

Tierney, Kathleen. 2014. *The social roots of risk: Producing disasters.* Stanford, CA: Stanford University Press.

Xerandy, Znati, Taieb, and Louise K. Comfort. 2016. "A cost-effective, environmentally aware undersea infrastructure to enhance community resilience to tsunamis." *Safety Science* 90: 84–96.

Chapter 5

Community-Based Shelters: Design, Construction, and Implementation

Febrin Anas Ismail and Abdul Hakam

In the first decades of the twenty-first century, a series of significant earthquakes struck the western part of Sumatra Island, Indonesia. Although many earthquakes have occurred in that area, experts still predict the seismic gap near the Mentawai Islands, which can cause a big earthquake, to generate a tsunami in the near future. Padang, the capital of West Sumatra Province, is located about 200 kilometers from the Mentawai Islands and has the potential to be inundated by such a tsunami. Approximately six hundred thousand residents of Padang live near the beach and are exposed to tsunami risk. To protect the people, the Indonesian government seeks to provide tsunami shelters in the exposed area. Due to a limited budget, however, only five shelters have been built, each with a capacity for two thousand to three thousand people (Kota Padang 2018). This number is far from sufficient for protecting the population at risk. To solve this problem, the Padang municipal government has invited the owners of multistory buildings to allow their buildings to function as shelters (Akbar 2017). This plan requires an assessment process to ensure that the prospective buildings are reliable. Unreliable buildings then need to be retrofitted to meet the safety standards for shelter. In fact, this expensive additional cost for retrofitting led building owners to reject the government proposal.

We propose a design for self-supported community-based shelters built and managed by community residents. These shelters would use existing public facilities such as mosques or *mushollas* (small mosques) that are close to the residents' houses. In Padang City, each sub-sub-district has at least ten mosques or mushollas within an area of 100 to 900 square meters (Kota Padang 2018). The shelters may be built based on the concept of upgrading the quality and quantity of the mosques.

The construction costs are based on the *infaq*, a type of charitable dona-
tion in Muslim terminology that is part of a worship service to God that
will be rewarded in the hereafter. With this concept, the construction
of shelters that require high costs and faster time to build can be done
hand in hand with the community's residents. This chapter presents the
structural design, construction methods, and implementation process for
building community-based shelters. These procedures may be adopted
as guidance for community residents who live near the beach to build
safe, reliable, and affordable tsunami shelters in Indonesia

Mitigating Tsunami Risk

The recent series of big earthquakes that have stricken the western part
of Sumatra Island, so far, have not triggered a tsunami, but experts still
predict that a big earthquake that can generate a tsunami is very likely to
happen in the near future (Anonim 2020). This risk is due to the so-called
seismic gap that exists near the Mentawai Islands. Padang City, located
about 200 kilometers from the Mentawai Islands, has the potential to
be destroyed by such a tsunami. This information is well understood by
stakeholders in the area, including government agencies, private sector
entities, nongovernmental organizations, and communities of residents.
A map indicating the areas exposed to tsunami risk is shown in figure 5-1.

Many tsunami mitigation efforts have been carried out so far (Hoppe
and Mahadiko 2009; Imamura 2009; Taubenböck et al. 2009; Muhari et
al. 2010; KOGAMI 2011; Imamura 2012; Spahn et al. 2012). One alternative is the so-called self-support shelter that is built and operated by the
community. It utilizes existing public facilities that are located in resi-
dential neighborhoods. In Padang City, each Rukun warga (a community
group of approximately one hundred to two hundred households) has at
least one musholla located in an area of approximately 100 to 400 square
meters and is easily found in Muslim communities (Sistem Informasi
Masjid 2020).

There are at least three advantages to using a mosque or musholla as
shelter. First, the lands are already available, so there is no need for an
additional budget to provide new land, which is reasonably expensive.
Second, mosques and mushollas are generally located close to community residents, so they can accommodate the surrounding community
residents as the tsunami evacuation shelters in time. Third, substantial
funds are needed to build shelters; therefore, the concept of upgrading

FIGURE 5-1. Tsunami-exposed area of Padang City. (Map created with data from many sources in a project launched by the mayor of Padang City, Mr. Fauzi Bahar, in 2010; https://pgis-sigap.blogspot.com/2011/03/peta-evakuasi-tsunami-kota-padang.html.)

the quality and capacity of a mosque or musholla may interest people in donating their money for religious reasons. Although it will require a long time to construct, using a step-by-step process depending on the incoming donations, a shelter surely can be completed. By using proper construction methods, the existing mosques and mushollas do not need to be destroyed before the shelters are completed.

During the Aceh tsunami in 2004 (Jatmiko 2014) and the Palu tsunami in 2018, mosques and mushollas were proven to withstand the tsunami's forces, as shown in figure 5-2. Although these mosques and mushollas were not designed based on the tsunami loads, the buildings survived due to their good-quality construction. Moreover, if new mushollas are designed based on tsunami loads, people will be convinced that these shelters will withstand tsunami forces even more reliably.

The procedures for community-based shelter development in terms of selection of sites, structural design, capacity and accessibility for the neighborhood residents, management, and construction methods are discussed here. The procedure may be used as guidance when planning and building community-based shelters in Indonesia.

Shelter Selection

Selection for a suitable location and building to be used as a shelter is the most important step in the tsunami evacuation procedure. The choice of tsunami shelter must basically be close to the location of the residents of the existing community who need shelter from the tsunami disaster. Therefore, the chosen location must be geographically in the

FIGURE 5-2. Mosques after tsunami: (left) Aceh, 2004 (Jatmiko 2014); (right) Palu, 2018 (Nourse 2018).

tsunami-prone area. Certain criteria, however, ensure that the selection of shelters is not only effective in saving the surrounding community residents, but also efficient in management and maintenance. These criteria must be determined in the selection of shelter locations so that the main functions of shelter during the disaster are fulfilled and the location can be used in conditions when no disaster occurs. To plan the evacuation process effectively, various factors must be considered in determining shelter location, as discussed below.

Distance from the beach

The distance of the building from the sea will greatly affect the strength of the shelter structure. The closer to the beach, the greater the tsunami force that must be restrained by the shelter. For this reason, the distance of shelters from the shoreline should be more than 100 meters. Moreover, the distance from the shore also determines the requirements for keeping the shelter safe from coastal dynamic mechanisms such as extreme waves, storms, and abrasion.

The location of the shelter is also psychologically decisive for people who will evacuate from the tsunami. People who evacuate themselves will feel that it is safer to go to a shelter away from the beach. Conversely, when evacuating from a tsunami hazard, people will feel endangered and doubtful if they need to move closer to the sea to reach shelters near the beach.

Local maintenance organization

Shelters for tsunami evacuation must be ready for use in case of a tsunami and must be maintained for immediate activation should a tsunami occur. To keep shelters in ready condition, a local organization to manage is necessary. Thus, a community-based, independent shelter must be located where there is a local maintenance organization that can easily manage it and be based in the community that will use it.

Good access

The tsunami evacuation shelter should be located where it can easily be accessed, both in normal conditions and in case of a tsunami. Access to the shelter should be via regular roads that are not specifically designated for evacuation. This ease of access also has advantages in terms of maintenance. In addition, these roads would be familiar to prospective shelter users, namely the surrounding community. Tsunami shelter access can be a public road that also facilitates evacuation for other people who are not part of the community, but who are incidentally in locations around the shelter when a tsunami threat occurs.

The number of routes of access provides a benefit to allow people to get to the shelter quickly. Further, the access capacity of the route, or width of the road, to the shelter should be in line with the number of people who will evacuate the neighborhood. In addition to the number and capacity of roads leading to the shelter, it is also important to avoid any obstacles en route. The existence of bridges or other structures that are likely to collapse during earthquakes and liquefaction areas that could topple buildings needs to be anticipated so that the evacuation process, when needed, can run smoothly and effectively (Comfort et al. 2009).

Land area

The chosen location for the development of the shelter must be large enough to accommodate the facility. The minimum area of land for a shelter must be sufficient to construct the building, as well as access roads and stairs to the tsunami-safe floor elevation. If possible, the land area should also accommodate other emergency facilities needed during a tsunami, such as a clean water supply and a logistics warehouse to store supplies for several days.

Soil conditions

The tsunami shelter must be located in an area where the soil is safe and stable against other hazards, such as earthquakes, liquefaction, abrasion, flooding, and sedimentation. When located in an area that is safe from other hazards, the shelter can function reliably in case of a tsunami. The security of the soil conditions for the shelter must be verified by good geotechnical investigation and soil stability analysis.

Selection method

To ensure a valid selection of shelter locations, a good selection procedure must be developed to meet the tsunami shelter criteria and produce the best choice. Various tsunami shelter selection models, such as those by Purbani and colleagues (2015) and Chang and Kim (2014), can be modified to serve the purpose of a community-based tsunami shelter.

Structural Design

The tsunami shelters must be structurally strong and be able to remain intact before and after being shaken by a big earthquake. Furthermore, the shelters must be designed to withstand the brunt of a tsunami and remain in safe condition being after hit. Overall, the shelters must have a structure that can defend against the loads that work on them.

Planned design

In addition to structural strength, tsunami evacuation buildings should serve multipurpose activities and require minimal building maintenance. Therefore, the tsunami shelters must be designed by professional planners who are capable of fulfilling the criteria. The selected planner must have sufficient competence to consider various tsunami shelter facilities and equipment properly. The planner must be able to analyze the shelter structure such that it is safe from the combination of applied working loads on it. The lesson learned from the failure of the tsunami shelter buildings in Lombok, Indonesia, in 2018 shows that shelters designed by an incompetent planner result in unsafe shelters against earthquake and tsunami.

Selection of structural formations

The structural formation of the tsunami shelter must take into account the forces that work against the building. To survive from heavy tsunami loads, the columns in the direction of the incoming tsunami must be made stronger than in the other directions. In the direction of the front part of the tsunami, the building's load should be calculated to include a minimum number of columns to avoid the accumulation of forces due to the tsunami. Columns should also be designed in a round shape or streamlined to minimize tsunami loads. The architectural component on the first floor where the tsunami load hits is made of lightweight material or made with holes so that it can flush the water smoothly. The architectural sections of the shelter in the direction of the tsunami can be designed to collapse at first if hit by tsunami water.

Selection of material

The main structure of the shelter must be made of a material that can exist for a long period of time (say, fifty years). Because the tsunami shelters are generally located near the sea, the material used must be resistant to corrosion or proven functional in salty environments. The main structure's material can be made of reinforced concrete with a concrete cover of about 5 centimeters. Stainless steel or steel with a stainless cover is also a good material that can be used for tsunami shelter buildings.

Soil investigation

A soil investigation is carried out to determine the site conditions of the shelter location and to analyze the requirements for the foundation. Soil investigations, such as boring tests or other tests, must be conducted in accordance with geotechnical code specified by the Indonesian National Standards Agency (Badan Standardisasi Nasional 2017). The results of a soil investigation will provide information about the soil types and layers, groundwater level, soil strength at a particular location. This information is a very important parameter for the purposes of carrying capacity and stability analysis of shelters. For sand soils with low groundwater levels that potentially liquefy, the shelter's foundation must be designed to guarantee that it will survive during liquefaction.

In the American Society of Civil Engineers/Structural Engineering Institute's manual (ASCE/SEI-7) for minimum design loads in building construction (Structural Engineering Institute 2016), the code requires that the geotechnical design and the foundation system consider the effects of scouring, pore pressure, soil-softening behavior, and liquefaction. In many cases, tsunami shelters may need a deep foundation system. Pile design also must consider increased demands due to down drag and additional lateral forces, as well as increased unbraced pile length due to scour. The depth of scour due to tsunami can be estimated by numerical analysis or an empirical formulation from past tsunami observations (Badan Standardisasi Nasional 2017). Uplift due to the overall buoyancy and overturning moments due to hydrodynamic and unbalanced hydrostatic forces must also be calculated in the foundation analysis. It is permitted to use a protective fill, slabs-on-grade, geotextiles, and reinforced earth systems to take these forces into account in the design of foundations.

Load system

The shelter structure must be designed based on the number of possible working loads that will act on the building during its lifetime. The loads that must be taken into account are *self-weight, life load, earthquake loads,* and *tsunami loads*. Self-weight is the gravity load of the entire building material, which is fixed for the lifetime period. Life load is any loading that is not fixed, depending on the code and special utilization of the building. Earthquake loads are the loads according to the code that has been determined for earthquake-safe building in Indonesia regulations. Tsunami loads are any forces that work due to tsunami waves and any materials in the wave. Tsunami loads, according to the U.S standard for designing structures for vertical evacuation from tsunamis, (FEMA 2019: P-646) include hydrostatic forces, buoyant forces, hydrodynamic forces, impulsive forces, debris impact forces, and debris duration forces.

Structural Analysis

The tsunami shelter structure analysis, as well as foundation analysis for the applied loads, should be conducted using numerical software. A number of software programs have been developed to obtain structural responses in the form of internal forces, displacement, stress, strain, and other parameters. Some software programs that can be used include SAP2000, ETABS, NASTRANS, and ANSYS. The results of the calculations are then used in the evaluation of structural elements so that they meet the criteria for strength and stability, as well as other requirements.

Construction Method

The construction stages necessary to turn a mosque or musholla into a shelter building may not interfere with, or at least minimize any disruption to, the worship activities in the existing building. It is vital to keep the mosque or musholla functioning as a worship place at all times, including during construction of the shelter. The existing mosque cannot be demolished before the completion of the replacement mosque as shelter. When the replacement mosque is completed, the columns in the middle of the designed new building must be minimized or eliminated. The construction process is likely to be time-consuming for the community-based shelter because the budget is gradually coming from the community itself. The construction process depends on the available budget. The stages of the construction process that meet the above concept are described below.

The first stage is constructing the foundation. The foundation is built around the existing mosque according to the predetermined column plan.

If the foundation, using piles, is constructed by placing the foundation with the boring system, not the hammer system, vibration and noise from the construction site will not bother people worshipping in the existing mosque (figure 5-3a). The construction of the columns can be started on the completely finished foundation. The construction of foundations and columns is only done around the mosque because there are still existing buildings in the middle of the span (figure 5-3b). After the columns are firmly erected, construction of the beams can proceed. Because there is no column in the middle of the building and the span between columns is quite long, the beams placed perpendicular to the likely front of the tsunami are used as the main beams to support the other beams. This supporting beam is made larger in capacity than other beams because it will carry the transferred loads from the other beams.

The supporting beams can be made of I-beams (figure 5-3c). The I-beam installation techniques can be utilized with lifting equipment such as cranes or rolling chains. If a truss structure is used, it can be assembled on the ground and then installed truss by truss on the column tips. The next step is the installation of the first floor and ramp stairs (figure 5-3d). After this stage, the existing building can be demolished to allow the construction of walls and other architectural components on the ground floor. It is estimated that when the construction budget has reached 50 percent of the final, the evacuation site can be used. The third step is the construction of the upper floors if the shelter capacity is still insufficient to accommodate the community (figure 5-3e and f).

Equipment used

The equipment used for the construction of community-based shelters must be good, workable, and affordable. The use of heavy construction machinery should be as minimal as possible to keep long-term costs low. The design requires efficient construction methods and materials but must also meet criteria for safety and sufficient strength.

Monitoring and supervision system

During the construction process, monitoring and supervision must be carried out carefully and properly to ensure that the shelter can be constructed smoothly and to avoid failure during the construction and utilization stages. Thus, assistance from experts is fundamental. The assistance effort can be carried out using a collaborative scheme with qualified university experts. This scheme has the benefit of reducing the cost of construction of the shelters while providing optimal results because it is carried out with the supervision of experts who concentrate on construction problems in these fields.

Shelter Evacuation System

Moving many people at the same time requires proper planning and measurable performance. In planning the evacuation process to the shelter, who does what, how, and when must be clearly designated. The plan needs to be prepared before a tsunami occurs. One evacuation model that has been developed and tested in the field is complementary to the concept of an independent shelter where evacuation will be carried out by the community residents who live near the shelter (Li and Xerandy 2019). This system principally organizes who will be the leader(s) and who are the followers. It also has equipment needed for interactive communication during the evacuation process. The existence of an independent communication system (Raspberry Pi network) is very important because it can operate when public electricity is not functioning. This tool is designed based on an Android operating system, which almost all people have the capacity to use via their mobile telephones. For intercommunity communication, a Raspberry Pi device is placed along evacuation routes within 200 meters, which allows users to maintain Wi-Fi connections to store and exchange information from leaders and followers in the community. The Raspberry Pi network is supported by a battery system that can function for an hour, which is longer than the estimated evacuation time of 30 minutes. Community leaders will be supported by a satellite mobile phone for communication to the emergency operations center and among the leaders, as presented in chapter 4.

Conclusion

The limited budget problem for shelter development was addressed by Padang City government's invitation to the owners of multistory buildings to allow their buildings to function as shelters. This effort requires an assessment process to ensure that prospective buildings are reliable, however. In fact, retrofitting unreliable buildings to meet the safety standards of shelter would require expensive additional costs, leading the building owners to reject the government proposals.

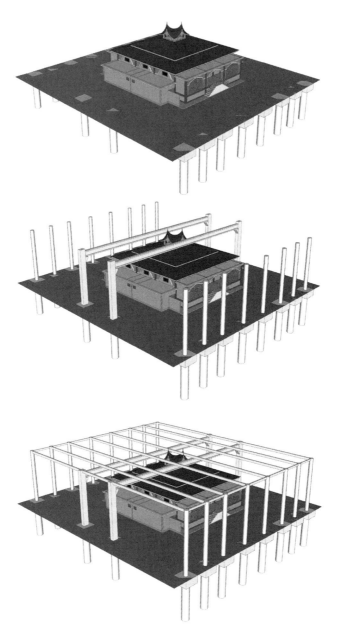

Figure 5-3. Stages of construction. See text for details.

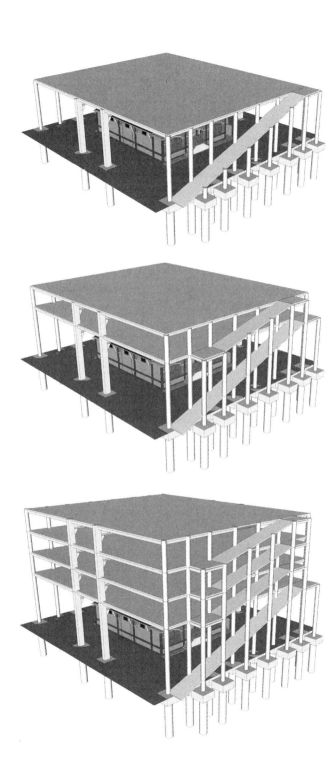

Community-based shelters are the next best solution at the moment. Community-based shelters are self-supported shelters that are built and managed by community residents. These shelters utilize existing mosques or mushollas (small mosques) located near the residents' houses. The construction of shelters that require high costs and time can be done hand in hand with community residents, but the development of community-based shelters has to meet the requirements for structural design, construction methods, and implementation processes. The procedures described in this chapter can be adopted for community residents in building community-based shelters in Indonesia and other nations with traditions of self-organization and commitment to reducing tsunami risk.

References

Akbar, R. 2017. "Dari 74 Shelter Pengungsian di Padang, Saparuhnya Tak Layak Pakai." *Oke-News*, February 3, 2017. https://news.okezone.com/read/2017/02/03/340/1608501/dari-74-shelter-pengungsian-di-padang-saparuhnya-tak-layak-pakai.

Anonim. 2020. "Expert talks about the potential of M 8.9 earthquake if a Mentawai megathrust occurs." *VOI*, November 13, 2020. https://voi.id/en/news/19675/expert-talks-about-the-potential-of-m-89-earthquake-if-a-mentawai-megathrust-occurs.

Badan Standardisasi Nasional. 2017. *Persyaratan Perancangan Geoteknik*, SNI. 8460-2017. Jakarta: Badan Standarisasi Nasional.

Bahar, Fauzi, Coordinator. 2010. Map, Tsunami Exposed Area of Padang City (Peta Evakuasi Tsunami Kota Padang). Collaborative mapping project coordinated by the Mayor of Padang City, initiated in 2010. https://pgis-sigap.blogspot.com/2011/03/peta-evakuasi-tsunami-kota-padang.html.

Chang, Jeong Keun, and Sung Gil Kim. 2014. "Proper Location of Disaster Shelters according to Evacuation Time-Focused on Coastal Areas in Hongseong Gun." *Journal of the Korean Society of Hazard Mitigation* 14, no. 1: 319–26.

Comfort, L., V. Cedillos, and H. Rahayu. 2009. "Planning Matters: Response Operations following the 30 September 2009 Sumatran Earthquake." American Geophysical Union, Fall Meeting 2009, December. Abstract id. U22B-04.

FEMA (Federal Emergency Management Agency). 2019. *Guidelines for Design of Structures for Vertical Evacuation from Tsunamis*, 3rd ed. FEMA P-646. Washington, DC: Applied Technology Council.

Hoppe, M., and H. S. Mahadiko. 2010. "30 minutes in Padang—lessons for tsunami early warning and preparedness from the earthquake on 30 September 2009." GTZ-GITEWS project publication. https://www.gitews.org/tsunami-kit/en/E4/further_resources/case_study/Case_Study_30_Minutes_in_the_City_of_Padang_30_September_2009.pdf.

Jatmiko, A. 2014. "How mosques survived the 2004 tsunami." *National News.com*. https://www.thenationalnews.com/world/asia/how-indonesian-mosques-survived-the-tsunami-1.636974.

Kota Padang (Padang City). 2018. *Padang City in Figures*. (Badan Pusat Statistik) https://padangkota.bps.go.id/publication/2018/08/16/49ae96955c1f8d70dcae6b52/kota-padang-dalam-angka-2018.html.

Imamura, Fumihiko. 2009. "Dissemination of Information and Evacuation Procedures in the 2004–2007 tsunamis, including the 2004 Indian Ocean." *Journal of Earthquake and Tsunami* 3, no. 2: 59–65.

Imamura, Fumihiko, Abdul Muhari, Erick Mas, Mulyo Harris Pradono, Joachim Post, and Megumi Sugimoto. 2012. "Tsunami disaster mitigation by integrating comprehensive countermeasures in Padang City, Indonesia." *Journal of Disaster Research* 7, no. 1: 48–64.

Muhari, Abdul, Fumihiko Imamura, Danny Hilman Natawidjaja, Subandono Diposaptono, Hamzah Latief, Joachim Post, and Febrin A. Ismail. 2010. "Tsunami mitigation efforts with pTA in West Sumatra Province, Indonesia." *Journal of Earthquake and Tsunami* 4, no. 4: 341–368.

Nourse, Jennifer. 2018. "The mosques that survived Palu's tsunami and what they mean." *RNS* (*Religion News*). https://religionnews.com/2018/10/16/the-mosques-that-survived-palus-tsunami-and-what-they-means/.

Purbani, Dini, Ardiansyah, Harris M. P., Hadiwijaya Lesmana Salim, Muhammad Ramdhan, Yulius, Joko Prihantono, Lestari Cendikia Dewi. 2015. "Determining Evacuation Route, The Temporary Evacuation (TES) with Its Capacity in Pariaman City by Using Geographic Information System (GIS)." Prosiding SIMPOSIUM NASIONAL MITIGASI BENCANA TSUNAMI 2015 TDMRCUniversitas Syiah Kuala didukung oleh USAID (PEER Cycle 3). December 21–22, 2015. No. ISSN: 2477-6440 Banda Aceh.

Sistem Informasi Masjid. 2020. "Data Masjid dan Musholla di Kota Padang." KEMENTERIAN Agama. Jakarta: Government of Indonesia. http://simas.kemenag.go.id/index.php/profil/masjid/page/?kabupaten_id=69.

Spahn, Harald, Michael Hoppe, H. D. Vidiarina, and Benny Usdianto. 2012. "Experience from three years development for tsunami early warning in Indonesia." *Journal of Disaster Research* 7, no. 1: 61.

Structural Engineering Institute. 2017. *ASCE 7-16 Minimum Design Loads and Associated Criteria for Buildings and Other Structures.* Reston, VA: American Society of Civil Engineers.

Taubenböck, Hannes, Nils Goseberg, Neysa Setiadi, Gregor Lämmel, Florian Moder, Martin Oczipka, Hubert Klüpfel, et al. 2009. "'Last-Mile' preparation for a potential disaster—Interdisciplinary approach towards tsunami early warning and an evacuation information system for the coastal city of Padang, Indonesia." *Natural Hazards and Earth System Sciences* 9, no. 4: 1509–28.

Chapter 6

Enabling Adaptive Collective Action for Communities at Risk: Responding to Tsunami Risk in Padang City, Indonesia

Yoon Ah Shin, Louise K. Comfort, Fuli Ai, and Febrin Anas Ismail

On December 23, 2018, an unexpected deadly tsunami originating in the Sunda Strait between West Java and the Sumatra Islands of Indonesia struck Carita Beach and other coastal towns, taking an estimated 430 lives and injuring at least 1,500 persons (Riadi and Griffiths 2018). The lack of timely warning to inform people to evacuate from the oncoming tsunami heightened the danger, again. After the tragedy of the 2004 tsunami that struck the city of Banda Aceh and the northern coast of Sumatra, disaster managers in Indonesia focused on building a technical tsunami detection and warning system. This focus misses a critical component of the tsunami early warning system: the human behavioral factor. The ultimate goal of a technical early warning system is to inform people about a coming risk in a vulnerable area and urge them to evacuate. The early warning system can be successful only when community residents recognize the warning alarm and immediately evacuate, taking actions supported by a successful mobilization of governmental adaptive plans for both governmental rescue agencies and community residents. We investigate this critical point of human cognitive and behavioral interaction in the continuing discussion of how to build a successful early warning system in Indonesia by highlighting the significance of collective action among community residents and governmental agencies.

For community evacuation, the National Disaster Management Agency (BNPB), established in 2008, has developed a standard procedure to make quick decisions, coordinate reactions, and communicate clear guidance. According to the current evacuation plan for a community, the Indonesian Meteorology, Climatology, and Geophysics Agency (BMKG) provides earthquake and tsunami information to BNPB and local governments immediately after sensing an undersea earthquake.

BMKG updates the earthquake threats to BNPB and local governments. Based on this information, local governments decide whether to call for an evacuation and provide guidance to a community at risk (Hoppe and Mahadiko 2010). When a local government decides that a community should be evacuated, the government alerts the community residents through public siren calls, media sources, and text messages to individual mobile phones. After the alerts have been sent, it is estimated that residents in a danger zone have 20 minutes to evacuate. Within those 20 minutes, however, there is a lack of specific operational guidance concerning how to evacuate a community (Hoppe and Mahadiko 2010). Personal capacity to make quick, appropriate decisions is essential.

We argue that community residents should take primary responsibility to evacuate themselves in this time-critical situation. To help community residents respond right away, it is critical to know to what extent tsunami early warnings are recognized by people at risk and whether the people are able to take an appropriate response. Our research focuses on ensuring that people receive an alert, recognize the risk in a dynamic situation, and act collectively to evacuate (Comfort 2007). Collective situational awareness serves as a backbone to support coordination, especially when multiple, diverse respondents from different locations need to collaborate in a fast-changing situation. Collective situational awareness leads these respondents to shape a comprehensive operational plan by understanding, adjusting, and coordinating their respective roles accordingly in response to an ongoing situation. To create collective situational awareness within and across respondents at different locations, such as decision makers at an emergency operational center (EOC) and community residents at the disaster site, continuous information flows need to be maintained through vertical and horizontal communications supported by resilient information technology (IT) communication infrastructure.

This chapter addresses the question: Does an increase in collective situational awareness through IT structure increase collective action in risk environments? To be specific, we address the 20-minute community evacuation plan by proposing to increase collective situational awareness among participating responders at the EOC and affected community residents through multidirectional communication. We developed a system dynamics simulation model to demonstrate the effectiveness of multidirectional communication on increasing collective situational awareness that, in turn, results in cognitive readiness for collective action to evacuate.

To support multidirectional communication across multiple respondents, we developed an IT communications infrastructure and tested its functionality in hypothetical situations (see details in chapters 3, 4, and 7). We used test results from the IT communications infrastructure, empirical data for cognitive effects, and demographic information from Padang City to create a system dynamics simulation model of evacuation. The model estimated the length of time for the entire population in Padang City to be cognitively ready to participate in a collective community evacuation supported by wireless IT communications infrastructure. By integrating the social and technical components of the information system and emphasizing the critical need for an innovative interdisciplinary approach, this analysis demonstrates the linkage between the system's technical function to communicate risk and the community's collective recognition of that risk that motivates people to take immediate action.

Collective Situational Awareness in a Tsunami Evacuation

Recognizing the threat of tsunami hazards is the first step that structures collective action for evacuation in communities exposed to tsunami risk. Once decision makers decide to evacuate community residents, it is crucial for all organizations participating in response to tsunami risk to gain a common picture of the ongoing threat and adapt their strategies to manage the situation. This section discusses the role of collective situational awareness in reducing cognitive gaps for participating organizations and engaging them successfully in multidirectional coordination.

Collective situational awareness

Situational awareness indicates a cognitive process that illustrates how people build a composite picture of a changing situation by combining new information with existing knowledge in working memory. Situational awareness supports people in making decisions for appropriate courses of action by predicting the future status of the situation (Fracker 1991). Situational awareness is highlighted as a significant cognitive factor required in complex, collaborative situations among multiple agents such as aviation and air traffic control domains (Endsley 1995; Klein 2017; Nofi 2000). In the fast-changing situation of disaster response, situational awareness also serves as a critical function enabling people to take immediate action by adjusting to a changing situation, because most people do not act until they have a holistic picture of the environment with appropriate comprehension of the relevant (and implicit) components and their relationships (Wellens 1993; Endsley 1995; Croft et al. 2004). When people are fully aware of the dynamic situation, they can project not only future movements in the environment, but also future actions of the system of disaster response operations.

Individuals, however, have different degrees of situational awareness for the same situation due to different mental models that are substantially influenced by personal experiences (Endsley 1995). These different mental models direct them to pay attention to, and capture different features from, the same situation. For example, community residents who have experienced a tsunami previously have a better idea of what to expect and how to protect themselves than those who have never experienced a tsunami, because the experienced people are more capable of integrating the environmental features, interpreting the meaning of the situation, and predicting possible changes in situational states.

Even if a person reaches a higher situational awareness level, individuals have different degrees of cognitive capacity to understand and predict the same situation, because each one's mental model for facing the context is unique. The discrepancy among individuals' cognition of risk undermines collaboration even on the same team, delaying time to save lives. For example, decision makers from the fire department at an EOC located in a city government could have different priorities and modes of community evacuation compared to firefighters at the disaster site, because the decision makers could miss the critical immediate context in a local neighborhood, such as an unexpected building collapse in a public area. Tsunami evacuation is a very complex situation, requiring comprehensive collaboration not only within, but also across, various response teams and community residents (Spahn et al. 2010).

We expand the concept of situational awareness within a specific team to multidirectional coordination in disaster operations for a whole community. In the tsunami evacuation plan for Padang City, the disaster response plan consists of multiple plans for response teams. Individual government agencies, such as the fire department and the police department, prepare to serve their organizational missions both at the EOC and in the field in response to a tsunami. Specifically, community leaders play an essential role as local leaders in neighborhoods by assisting in the evacuation of neighborhood residents. They also collaborate with responsible government agencies to evacuate neighborhood residents when those agencies arrive at the site. To serve this role, the neighborhood leaders should be fully aware of how the city EOC allocates evacuation routes and shelters and how the city EOC adjusts its plans to the current situation.

The city EOC also needs to be aware of how the neighborhood evacuation is progressing so that the incident commander can allocate the right resources in the right amount to the right place to meet the timely, urgent needs in that neighborhood. The quality of coordination between the leaders and city EOC informs the community residents' quick evacuating action by providing a clear picture of the evacuation process that adapts to the imminent risk. This continuous process for understanding adaptive changes in response plans should occur not only at individual and team levels, but also at the collective cognitive level. Collective situational awareness is a comprehensive concept that indicates neighborhood leaders, response teams from the city EOC, and residents of the affected community are ready for collective action in disaster response.

How can collective situational awareness be improved?

Effective communication among individuals, between outside teams, and among outside and inside teams are prerequisites for achieving a high level of collective situational awareness for a community at risk. Communication as an action of vital information exchange provides input for building collective situational awareness. After individuals in a team have attained their respective levels of situational awareness, communication among team members increases to a common team understanding that strengthens coordination (Lloyd and Alston 2003; Nowell and Steelman 2015). Sharing each team's preexisting relevant knowledge, new environmental information, and expectations across various response teams lays a firm foundation to build situational understanding collectively by enlarging individual and team cognitive capacities.

In response to a tsunami threat, updating operational progress concerning the mission objective, individual tasks, and roles within a team's operation increases the collective capability of the whole operational system. In turn, this system-wide capability substantially increases collective situational awareness among participating respondents in an evacuation operation. Understanding individual tasks embedded in a larger operational picture allows individual respondents to tailor their behaviors more appropriately within the system. Thus, communication as a collective adaptive process results in not only acquiring, but also maintaining, a common operating picture among participants in changing situations (Nowell et al. 2018). Further, communication during tsunami evacuation strengthens all engaged participants in their capacity to create shared meaning, and this shared meaning, in turn, builds cooperative relationships among the actors (Klimoski and Mohammed 1994; Van Kleef and Fischer 2016). This process links shared understanding and knowledge of a situation with team performance, representing a collective coordination process. Therefore, multidirectional communication among diverse participants serves as an index of shared awareness to enhance coordination (Endsley 1995).

Bowtie Communication Principles during a Tsunami Evacuation

A complete framing of a tsunami threat should come not only from the EOC, but also from the affected community. The capacity of the city EOC to produce a realistic but effective evacuation plan depends on how aware the decision makers at the city EOC are of the ongoing situation in the threatened community. Such reflective evacuation plans, in turn, enhance the collective awareness of community leaders and residents. Developing collective situational awareness requires both horizontal and vertical communication among diverse respondents, including community residents. The tsunami evacuation plan of the Indonesian government from the time of the 2004 tsunami until now has mobilized a unilateral, top-down communication model, using channels such as television, radio, mobile text message, email, fax, and public sirens from the EOC to the public (see chapter 2). Such top-down communication flow results in delayed evacuation because people do not grasp the real-time conditions. Heavy cognitive loads tend to undermine a person's ability to figure out the situation and determine what to do under tight time constraints (Mayer and Fiorella 2021). During the evacuation period of the 2004 tsunami, 67 percent of the residents in Padang City tried to reach out to family, friends, and neighbors to get more trustworthy and practical evacuation information (Birkmann et al. 2008). Understanding communication principles and models that support the most effective evacuation process is critical to future adaptation of the Indonesian evacuation plan.

The bowtie model proposed by Comfort and colleagues (Comfort 2005; Comfort et al. 2010; see also chapter 1) addresses the multidirectional communication required in tsunami evacuation. The modified

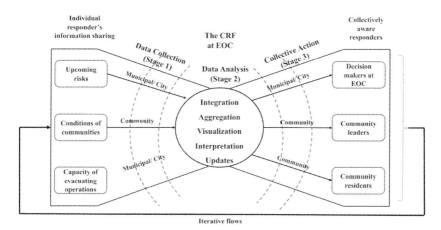

FIGURE 6-1. Modified bowtie architecture. (Original model from Comfort 2005.)

bowtie model in figure 6-1 focuses on a tsunami evacuation situation and identifies critical information required to support the evacuation process and each respondent's role in the Indonesian context. This model consists of three stages to drive collective evacuation by participating respondents: collection of critical information (stage 1), data analysis (stage 2), and collective action (stage 3). All critical information concerning community evacuation is collected via an interagency communication platform called the community resilience framework (CRF) at the city EOC (stage 2 in figure 6-1). The CRF is a central coordinating portal site where all relevant agencies have access to real-time input. The CRF infrastructure acts as a key to the bowtie model of multidirectional communication by providing a profile of changing risk for the whole community and transmitting relevant information action agencies to support decision making regarding evacuation in near real time.

According to the Indonesian government's disaster plan, when the signals of an undersea earthquake or landslide and corresponding change in the water column are captured by an undersea sensor and transmitted to the national scientific agency, BMKG, for validation, BMKG sends tsunami alerts and earthquake information to BNPB, local disaster management agencies (BPBD), and municipal and regency EOCs (see chapters 2 and 3). When decision makers at a local government's EOC decide to evacuate residents based on estimated earthquake and tsunami impacts, individual response teams or organizations need to update their situational information, such as current availability of personnel, availability of assigned evacuation routes and shelters in each community, and availability of community leaders to a municipal or regency EOC (Ai et al. 2016).

The CRF decision support system collects incoming information about an imminent threat and aggregates, integrates, and analyzes the incoming data with existing knowledge to produce updated information that is relevant to each action agency. This computational data analysis substantially contributes to reducing cognitive energy loads and saves time for decision makers at the EOC. Also, this process minimizes individual misperception about the situation caused by limited human cognitive capacity. When individual responders, including decision makers at the municipal EOC and community leaders and residents in the neighborhoods, receive timely, valid information regarding evacuation operations that reflect the current context, they are able to build up a collective operational picture across different locations of operations. This collective situational awareness enables them to structure and adjust their immediate tasks and actions according to the system's response operations (stage 3 in figure 6-1). Iterative communication flows among participating respondents as designed in the bowtie model significantly enhance collective situational awareness. Even if each respondent perceives the situation differently due to individual situational awareness discrepancy in the first stage of the operation (Hale et al. 2005), an iterative process of communication, including error detection and correction, ultimately reduces gaps in perception among individual respondents acting in altered conditions (Comfort 1999).

Practical Application of the Bowtie Principle: Multidirectional Communication Flows

To implement the bowtie communication model to support collective evacuation actions in the Indonesian context, we identify critical actors, types of information, and information flows to deliver multidirectional communication. Designing the direction and channels of communications between operations personnel, community leaders, and community residents *before* a tsunami occurs increases the timeliness and clarity of information that enables collective action in an actual event. This section provides an overview of the communication flow among community actors engaged in multidirectional communication to coordinate timely evacuation.

Critical actors for evacuation communication

Effective evacuation depends on timely, informed communication among three major groups in a community at risk: (1) operational personnel with legal responsibility for protecting community residents from harm, (2) community leaders in the neighborhoods who act as intermediaries between operational personnel and residents, and (3) residents of the community who are at risk. Depending on their roles and locations, different actors search for—and exchange—varying types of information in the collective decision process. Determining the communication pathways within and among these three major groups is vital for collective action under urgent conditions.

Emergency personnel at the municipal EOC. Emergency personnel at the EOC have both expertise and authority to make official decisions and provide planned evacuation scenarios to the public to practice plausible strategies during a crisis. They are regularly trained to analyze risk, develop mitigation plans and strategies, and implement evacuation drills. When they receive a tsunami alert from BMKG, BNPB, and BPBD, they interpret that risk for the city and, based on previous knowledge and experience, decide which neighborhoods should evacuate. Even with expert knowledge and experience in disaster response, emergency personnel cope with multiple tasks simultaneously, including not only community evacuation but also coordination with other government agencies. To process real-time information from an affected neighborhood, they need automated computational support for data collection, analysis, updates, and dissemination to enhance collective neighborhood evacuation.

Community leaders. Community residents have a strong interdependency, especially when they confront a common threat, strengthened by a sense of attachment to people within the same physical place. Such a sense of communal attachment drives them to recognize a crisis as "our" issues to resolve together with other members in the same neighborhood (Lyons et al. 1998; Comfort 1999; Bellifemine et al. 2005). Community leaders acknowledged by community residents are familiar with the local context, culture, and people. These familiarities help the community leaders increase collective situational awareness of the residents effectively because they use local and common languages to update the current situation and give specific guidance to the residents for an evacuation

(Mileti 1999). Furthermore, collective evacuation action among community leaders in neighboring subdistricts serves a critical role to meet immediate needs in a community. Because leaders in neighboring subdistricts share the same training experience, which binds them with the common goal to save "our" communities, they can effectively share resources and adapt their actions to one another's needs (Ostrom 2005). Even if their shared experiences and communication miss a critical picture of the evacuation operation due to cognitive distress and individual discrepancies in mental models, they can overcome such challenges by improvising temporary rules which support, retain, or revise collective situational awareness during an evacuation (North 2005).

Community residents. Community residents play a proactive role in assisting their community leaders as first responders during the tsunami evacuation. Community residents substantially contribute to collecting real-time information and providing the community leaders with updates concerning the current physical and demographic conditions of the community. Such updates are critical for mobilizing adaptive collective evacuation. Scientific knowledge and official information provided by the EOC are often placed in conflict with residents who want immediate action, yet local knowledge is necessary for assessing the true state of affairs of the physical community—regarding both its people and the geographical area—with more accuracy, detail, and understanding of potential consequences of risk (Kostoulas et al. 2008). Shared local knowledge, languages, and norms among leaders and community residents help them communicate effectively under high stress in an altered world and, in turn, promote collective situational awareness to evacuate the community.

Five pathways of information flows in collective tsunami evacuation

Of the multiple pathways for information to flow within a community, five are critical for effective activation of community residents under the urgent stress of a tsunami alert. They include the following pathways, from decision makers to community leaders and vice versa.

From decision makers at the city EOC to community leaders. When the municipal EOC receives a tsunami alert and information from BMKG and BNPB, the decision makers at the municipal EOC analyze the tsunami risk and its consequences based on multiple sources of information displayed on the CRF. To minimize the cognitive loads of these decision makers, the CRF's computational function analyzes the current risks in terms of evacuation guidance: degree of seismic movement, level of potential consequences in each community, and conditions of assigned evacuation roads and shelters. When EOC personnel decide to evacuate a vulnerable community based on current information, the CRF automatically disseminates tsunami alerts to community leaders in each neighborhood of the city with updated evacuation guidance, including assigned evacuation routes and shelters.

From leaders to community residents. The community leaders warn their assigned neighborhood residents to evacuate immediately by sending a tsunami alert with specific information for an assigned shelter and its evacuation path. This information indicates what actions the neighborhood residents need to take right away; otherwise, the residents would spend extra time to find detailed information to comprehend the exact situation and to find appropriate actions to avoid tsunami risk. In the proposed community information network, this information is visualized on a neighborhood map with familiar landmarks on the neighborhood residents' smartphones. This visualized information shows assigned shelters and evacuation routes and enhances neighborhood residents' situational awareness retrieved from long-term memory. This visualized information substantially helps those who have less knowledge or literacy to understand, such as young children.

From community residents to leader. During evacuation, neighborhood residents can communicate with their leaders to share risk information and request emergency help. Updating risk information and emergency requests from an affected neighborhood is critical to mobilize resources efficiently, minimizing further damage. The communications system consists of an application on a mobile phone with default icon buttons related to specific situations that community residents can easily use; this design saves time and energy during evacuation. For example, if a resident pushes a risk report button to inform the community leader about a collapsed bridge on an evacuation route, the leader receives detailed information concerning the sender's name and the location of the collapsed bridge. If the leader receives the same risk report multiple times, the leader recognizes the risk for the current evacuation route. Updates of the current situation from residents help a leader adapt the evacuation strategy in real time, minimizing further risks and damage in the neighborhood.

From leaders to other leaders in neighboring subdistricts. During an evacuation, each subdistrict could have a different level of damage based on its location in reference to the earthquake and tsunami risk. Whereas some parts of the community may be severely destroyed by collapsed buildings caused by an earthquake, other parts with low-rise buildings could be relatively safe. Through communication, community leaders can help one another by sharing specific information about their respective subdistricts, such as the exact location of an obstacle and specific type of resources needed, with other neighboring leaders.

From community leaders to EOC decision makers. Neighborhood residents play an important role in updating contextual, real-time information to increase predictability for further evacuation plans. Due to limited lack of experience and expert knowledge, they may report less significant information to leaders. Leaders need to screen information critical to the evacuation plan for the entire response system. For example, leaders do not request an evacuation route detour unless they receive multiple risk reports, including bridge and building collapse, fire, and other risks, from different neighborhood residents for the same site. When requesting a new evacuation route from the EOC, a leader sends information for the exact location of the current evacuation route and the type of imminent risks. The CRF at the EOC displays this request, analyzes the current situation, and updates a new evacuation route. This new information is confirmed by operational personnel at the EOC and sent to the requesting community leader. Table 6-1 illustrates the interconnectedness among the five paths of information flow that would support collective action in evacuation in practice.

Research Design, Measurements, and Data

This research tests the extent to which multidirectional communication among EOC, community leaders, and neighborhood residents enhances the residents' cognitive readiness by raising their collective situational awareness. Collective situational awareness increases the residents' readiness to mobilize collective action for evacuation while decreasing

TABLE 6-1. Five information pathways of communications

Communication Path	Purpose	Details	Bowtie Principle
From EOC decision makers to leaders	Initiation of evacuation	Send tsunami alert and (updated) evacuation route	Collective action
From leaders to neighborhood residents	Mobilize community collective evacuation	Send tsunami alert, information about evacuation routes, and locations of assigned shelters	Collective action
From neighborhood residents to leaders	Provide timely information	Report new risks in neighborhood; send messages for evacuation assistance	Data collection Data analysis
Between leaders in neighboring subdistricts	Assist neighboring leaders by sharing resources	Request *immediate* evacuation assistance when a leader is limited in movement or in resources	Data collection Collective action
From leaders to decision makers at EOC	Request updated evacuation routes and shelters	Request updated evacuation routes when current evacuation routes are not available due to (potential) dangers on the way	Data collection Data analysis

cognitive loads. The research design includes both social and technical perspectives. The social perspective of this model illustrates (1) how multidirectional communication flows increase collective situational awareness for the entire system of participating responders and (2) how improved collective situational awareness increases cognitive readiness to act collectively to evacuate while decreasing cognitive loads caused by time constraints. The success of multidirectional communication is essentially determined by effective performance of the IT communications infrastructure. If the public telecommunication infrastructure fails, communication is limited. Specifically, the frequency and speed of communication among the participating residents are determined by the success rate of computational information analysis and the stability of a network connection to transmit updated information among residents.

To demonstrate the interdependency between social and technical perspectives in the model of collective tsunami evacuation, we have developed a system dynamics simulation model using the AnyLogic program (AnyLogic 2018). A system dynamics (SD) model represents the nonlinear behavior of complex systems over time by setting up both static and dynamic variables (Borshchev and Filippov 2004). Unlike a statistical model, the SD model captures the interdependencies among multiple variables; when the impact of one variable increases, the impacts of relevant variables also are changed, either increasing or decreasing. SD models present the interdependency of cognitive factors engaged in individual and group decision-making situations under uncertainty (Gary et al. 2008). The collective evacuation is driven by a complex iterative process between human cognitive factors and communication behaviors supported by a technical system. Therefore, the SD model is a very appropriate method to test the effectiveness of our suggested collective evacuation model, capturing the complex, interdependent relationships among social and technical factors.

Simulation model design

This simulation model shows the flow of how the cognitive status of neighborhood residents would change from an individual inaction gap (IIG) into cognitive readiness for collective action (CRCA). The simulation starts with neighborhood residents who are in an IIG regarding evacuation, which indicates that those residents have not heard or have not recognized significant information of a coming tsunami. The stock level of CRCA increases when neighborhood residents who stayed in IIG become fully aware of, and are ready to join, an ongoing tsunami evacuation, which causes a decrease in the stock level of IIG. Once they recognize a tsunami threat, their cognitive loads increase from stress as they seek appropriate actions within a limited time (Galy and Mélan 2015). When they are collectively aware of the immediate action concerning what to do and where to go, they are cognitively ready to join a collective evacuation. Therefore, their readiness to act is determined by the degree of cognitive load and level of collective situational awareness. When the level of collective situational awareness overtakes the level of cognitive load, individual neighborhood residents are ready to join a collective evacuation.

The collective situational awareness depends on the success rate of the community leaders' communication with other participants to update information to the city EOC, neighboring community leaders, and their community residents. Termed SRLIT (success rate of leaders' information transmission), this rate is measured by the number of messages transmitted by community leaders through the neighborhood wireless network that are received by the intended recipients. As the gatekeepers of updated information in real time, the community leaders play a role of human central control to receive and transmit appropriate information to meet neighborhood residents' requests, which ultimately increases the residents' situational awareness. As shown in figure 6-1, community leaders collect updated information coming from the CRF, neighborhood leaders, and community residents. The performance of community leaders in sharing updated information with neighboring leaders and their residents substantially relies on the performance of the neighborhood wireless network (see chapters 3 and 4). First, the flow of updated information from the CRF to the neighborhood network depends on the CRF's capacity for timely data collection, aggregation, and analysis. Second, updated information exchange among community leaders and neighborhood residents occurs only when the Pi network successfully connects with each person's mobile phone at the same time.

Currently, there is no standard IT communications infrastructure that supports real-time multidirectional communications when the landline and internet networks are disrupted. It is necessary to develop a new community-based network system to support essential communication flows among residents operating in vulnerable conditions. This community-based network includes the Raspberry Pi network as well as the CRF (see chapters 3 and 4). The Raspberry Pi network plays a critical role in providing alternative internet access when landline or regular internet services are disrupted to support information flows among individual mobile phones and the CRF (see details in chapter 7). To check the capacity of the community-based network, we performed two field tests in Padang City, in 2015 and 2016, to examine the capacity of the CRF's computational data analysis and the network connections of the Raspberry Pi network. By installing ten Raspberry Pis with five mobile phones that synchronize with the CRF, we examined (1) how many phones can be steadily connected and exchange data among mobile phones and the CRF through the Pi network at a time and (2) how successfully the CRF analyzes and updates incoming information. After two field tests, we increased the capacity of the Raspberry Pi network to capture a communication signal even during dynamic movements, such as people running. Later, in July 2018, we tested the latest design of the land-based network on the University of Pittsburgh campus. We have tested ten Raspberry Pi devices with simulated one hundred mobile phones to simulate the information flow through the whole community flow during an evacuation. The final results are applied as measures of the success of multidirectional communication. Figure 6-2 illustrates the information flow that initially drops under stress of an extreme event and regains readiness for action through increased situational awareness, as well as the critical role of community leaders in transmitting information directly to the residents in their respective neighborhoods.

Measurements and data

We apply this model to predict the evacuation readiness time in Padang City. Therefore, we use the demographic information of five villages—Belakang Tangsi, Olo, Kampung Pondok, Kampung Jao, and Padang Pasir—in the Padang Barat subdistrict, Padang City: 21,054 community residents (BPSKP 2013) with ninety-nine volunteer evacuation leaders (Rahayu 2016). The rate of stock change (adaptive rate) is determined by the levels of collective situational awareness and cognitive loads. Collective situational awareness increases an adaptive rate to turn from IIG to CRCA by 0.224 ($p < 0.001$) (Setiadi and Birkmann 2010, referred in Taubenböck et al. 2013), whereas high cognitive load undermines the adaptive rate. In contrast, collective situational awareness decreases cognitive loads, which is increased by 0.206 ($p < 0.001$) because of time

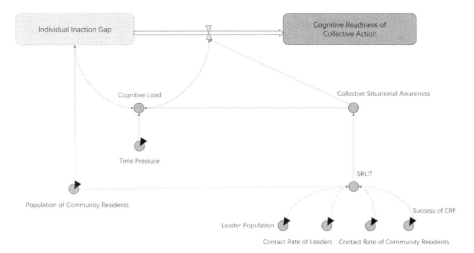

FIGURE 6-2. System dynamics model of collective tsunami evacuation.

constraints (Galy and Mélan 2015). Collective situational awareness is determined by the leaders' information transmission (SRLIT). To what extent could leaders' information sharing increase the collective situational awareness among the participating respondents? Fransen and colleagues (2011) show that during computer-supported collaborative exercise, information sharing effectively aligns goals and tasks among team members by 0.264 ($p < 0.05$).

Furthermore, the SRLIT is determined by the functionality of the neighborhood-based electronic network infrastructure specifically with three factors: (1) success rate of CRF at the municipal EOC, (2) successful communication rate with the assigned community residents (number of assigned community residents × success rate of Pi network connection), and (3) successful communication rate with neighboring leaders (number of neighboring leaders × success rate of Pi network connection). We assigned test results from the 2018 July field test held at the University of Pittsburgh and found that our Pi network connected multiple phones at a time with a success rate between 57 and 99 percent. This success rate was substantially influenced by installed locations, people's moving speed, and data loads on the Pi network. The success rate of data collection and analysis for the CRF was steadily around 55.09 percent (see details of the experiment in chapter 7). Detailed information on the variables included in this model is summarized in table 6-2.

Data analysis and results

This analysis estimates the time it would take to increase the cognitive readiness of community residents such as those in Padang City to join in collective evacuation supported by multidirectional communication. The analysis focused on determining the difference in time saved when running the electronic network—Pi network and CRF—at both maximum and minimum capacity. The expected time saved under both optimal and marginal conditions of the electronic network can be compared to actual evacuation time in Padang City. This capacity is represented as the potential time range needed for the entire community of residents to be cognitively ready.

The first scenario of the simulation is when each communication component is operated with the maximum capacity: 99 percent of success rate of Pi network and 55 percent of success rate of CRF. Figure 6-3 shows the simulation results indicating how long it takes for the entire community of residents to convert from IIG to CRCA. The CRCA

TABLE 6-2. Variables and measurements of collective tsunami evacuation model

Types	Variables	Definition	Details	References
Stock	Individual inaction gap (IIG)	The number of community residents unaware of an upcoming emergency situation	Population of community residents (PCR)	—
Stock	Cognitive readiness of collective action (CRCA)	The number of community residents who are cognitively ready to join a collective evacuating action	Inflowing number of PCR	—
Flow	Adoptive rate	The status changing rate from IIG to CRCA	$0.224 \times$ CSA/cognitive load	Setiadi and Birkmann 2010
Dynamic variable	Cognitive load	The consumption of heavy mental energy	$0.206 \times$ IIG/CSA	Galy and Mélan 2015
Dynamic variable	Collective situational awareness (CSA)	Collective development of composite picture of the situation by combining new information from multiple and diverse teams with existing knowledge in working memory	$0.264 \times$ SRLIT	Fransen et al. 2011
Dynamic variable	Success rate of leaders' communication with other responders (SRLIT)	Leaders' vital information exchange with decision makers at emergency operations center, assigned community residents, and neighboring leaders	(Leader population $\times$ CRL) + (PCR $\times$ CRCR) + SRCRF	—
Parameter	Population of community residents	The number of residents in a community	21,054 for five villages in Padang Barat subdistrict in Padang City	BPSKP 2013

Types	Variables	Definition	Details	References
Parameter	Leader population	The number of volunteer leaders to serve a community-evacuating process in a community	99 for five villages in Padang Barat subdistrict in Padang City	Rahayu 2016
Parameter	Contact rate of leaders	The success rate of Pi network to capture signals of nearby cell phones	Scenario 1: 0.99 Scenario 2: 0.57	Ai et al. 2023; chapter 7
Parameter	Contact rate of community residents (CRCR)	The success rate of Pi network to capture signals of nearby cell phones	Scenario 1: 0.99 Scenario 2: 0.57	Ai et al. 2023; chapter 7
Parameter	Success rate of community resilience framework (SRCRF)	The success rate for CC to update new evacuation guideline information after a data process (data collection, integration, and analysis)	0.55	Ai et al. 2023; chapter 7

steadily increases until the 72nd second, which is the turning point marking when the number of residents in CRCA status overtakes the number of residents in IIG status. Right after the turning point at the 72nd second, the CRCA exponentially increases. Within 24 seconds from the 73rd second to the 96th second, another half of the resident population who have stayed in IIG status are rapidly converted into CRCA status. This exponential increase is caused by the enhanced collective situational awareness which decreases the cognitive load level. At the beginning of the evacuation process, the multidirectional communication has not actively mobilized and does not make a big impact on increasing collective situational awareness among the participating residents. The cognitive load caused by time pressure plays a dominant role in determining the speed of converting awareness from IIG to CRCA. As the level of collective cognitive awareness keeps rising through the continuous multidirectional communication among decision makers at the EOC, community leaders, and residents, the impact of the collective situational awareness on decreasing the cognitive load gets stronger, which, in turn, effectively shortens the time of conversion. Figure 6-3 illustrates the point at which the conversion from collective cognitive awareness of risk to collective action occurs in a community under threat.

The second simulation scenario is preconditioned to operate with the minimum capacity of our IT communications infrastructure: 57 percent success rate of the Pi network connection and 55 percent success rate of the CRF's data analysis. The simulation results shown in figure 6-4 indicate that the turning point arrives much later than the first simulation results. The number of CRCA starts to overtake the number of IIG at the 217th second, and the entire community of residents reach to CRCA status around the 311th second. Compared to the first simulation, the second simulation takes three times and four times longer to reach a turning point and a completing point of status change from IIG to CRCA, respectively, as illustrated in figure 6-4.

In an interesting finding from the second simulation, we captured the marginal capacity, or threshold, of the communication system. The threshold of the system is a tipping point when the system operation suddenly or rapidly shifts to a new state so that the "system experiences a qualitative change, mostly in an abrupt and discontinuous way" (Jax 2014, 1). Thresholds of change occur mainly when a stable condition or the resilience of a system is broken (Anderson et al. 2009). If an external driver such as temperature changes gradually, an ecosystem state may respond gradually itself. In this case, there are no tipping points that show

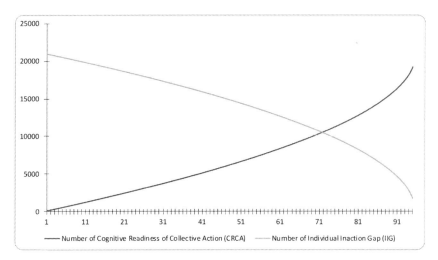

FIGURE 6-3. Analysis with the maximum capacity of IT communication infrastructure.

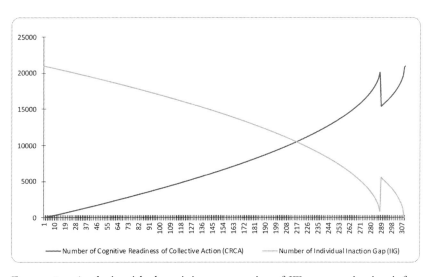

FIGURE 6-4. Analysis with the minimum capacity of IT communication infrastructure.

unexpected patterns of change in properties. In other cases, the response changes abruptly even though the external variable changes only gradually. Although there is little response change below a certain threshold value of the system, a small change in the environmental condition will cause an unexpected result once that threshold is crossed (Jax 2014).

A decision-making threshold is the value of the decision-making variable when the decision is made such that an action is selected or a commitment to one alternative is made, thus marking the end of accumulation of information. The tipping point of our system is at the 288th to 289th second, when the number of residents in CRCA condition suddenly decreases while the number of residents in IIG condition suddenly increases. To be specific, 4,579 residents who already reached CRCA status turn back to IIG status. The cognitive threshold occurs because of different levels of individual cognitive capacities. Limited human cognitive capacities cause biased judgments in one's capacity for error detection and correction; when people use their cognitive energy heavily for a long time, their cognitive function drops in decision processes in complex, uncertain environments (Simon 1996). Furthermore, individuals adopt varied strategies that promote error checking to different degrees, but, ironically, this error-checking strategy could be what makes a mistake (Jackson et al. 2016). This result shows that decision making in a fast-changing environment during an evacuation process requires extensive cognitive energy to construct actions, to monitor, and to control the outcomes. When one's cognitive capacity reaches a threshold, or limit of resilient function, a person may make a wrong interpretation of the updated information and a wrong decision. Therefore, if participating residents require a longer time to determine what is going on and what to do because of large amounts of new, unfamiliar information or because of late arrival of updated information, the multidirectional communication would result in distracting them from building up an accurate and comprehensive operational picture reflecting the changing situation.

These two results indicate that the capacity of the electronic neighborhood network plays a key role in enhancing situational awareness by decreasing cognitive load through vertical and horizontal communications among the residents and thus minimizing critical delays. Access to information is a critical factor to help people become aware of the threat, which drives immediate evacuation (Riad et al. 1999). Many failures of immediate response in disaster situations, including the Indonesian tsunamis, have come from the breakdown in communication infrastructures, which limits access to updated information (Taubenböck et al. 2013; Comfort 2007). The proposed communication platform works reliably only when both technical and social parts function effectively; if either part fails, the entire system slows down or fails. This simulation model integrates nonlinear relationships among social and technical factors, reflecting realistic human cognitive and behavior characteristics.

Conclusion

Findings from the simulations reveal the importance of an interdisciplinary approach for integrating human cognitive and technical components to build a successful tsunami early warning and community evacuation system. To address the question of whether collective situational awareness, supported by multidirectional communications, enhances collective action to reduce tsunami risk, we developed a system dynamics model to show that nonlinear interactions between cognitive and technical factors are critical during a simulated community-wide tsunami evacuation. The most valuable finding in our simulation signifies the capacity of the entire system to drive collective action in evacuating a community at risk from tsunamis, which cannot be captured in a linear-based statistics approach. Even though it is commonly believed that communication plays a critical role in increasing collective situational awareness, the result of our dynamics model emphasizes that prolonged communication could rather decrease individual cognitive capacity and cause cognitive errors in comprehending the situation and making informed decisions.

The operating capacity of the IT communications infrastructure determines the effectiveness of communication. First, high performance of the IT communications infrastructure quickens the time of the turning point of community residents from IIG to CRCA conditions by speeding up information transmission. Reaching the turning point more quickly plays an essential role in shortening the length of time to convert the entire population into CRCA condition. Another critical factor is the sustainability of the communications infrastructure to keep steady and fast information flows. During a community tsunami evacuation, environmental conditions change dynamically. Late arrival of updated information would not reflect the current situation, which increases the cognitive loads for people from time pressure. If this miscommunication is repeated over and over, people eventually would not follow the updated instruction because of mistrust, which, in turn, breaks down the entire system. Only a resilient, sustainable communication infrastructure to support timely multidirectional communications can initiate and enhance participants' cognitive readiness to drive successful collective action.

Supported by information flows, community leaders serve a critical role as information hubs to reduce coordination distances among the participants in the communications chain. Although individual responders are located in different places, the perception gaps about the situation among the multilevel responders is narrowed through critical updated information sent by the community leaders. Therefore, vertical and horizontal information flows through the community leaders, who serve as hubs, are central to connecting all responders in evacuating to the operation's central location. It is critical for community leaders to use discerning judgment to screen and categorize incoming information so that they share only timely, significant information with relevant responders without breaching the threshold of the residents' cognitive capacity.

In our analysis, reaching cognitive readiness for collective action takes far longer with the minimum capacity compared to the maximum capacity supported by IT communications infrastructure. In fact, even the worst scenario shows that it takes only 5 minutes for the entire population to comprehend the ongoing situation, with individual roles and actions corresponding to the entire operational picture; they are ready to take collective evacuation. Previous research focusing on the tsunami evacuation process in Padang City showed that, in general, people take 10 to 15 minutes to check the validity of tsunami information, gather more detailed information before proceeding to action, and assess how other family members and neighbors are doing (Hoppe and Mahadiko 2010). Through these information-search activities, people increase their cognitive loads as they seek to comprehend the situation and find appropriate action. When they start to take action or are on the way to a shelter, the tsunami has already reached a neighborhood located along the seashore. In this sense, the proposed communication evacuation model substantially increases the survival rate of neighborhood residents by saving critical time to evacuate to a shelter.

Further research could measure directly the cognitive impact on people's collective evacuation action by implementing a field experiment in Indonesia. Because of time and physical limitations, we applied empirical analysis results from field experiments and surveys in other research investigating people's cognitive and behavioral characteristics in a complex decision-making situation. Those results provide appropriate alternative data to support the arguments and develop a simulation model. A field experiment with neighborhood residents in Padang City using the IT infrastructure in an evacuation scenario would validate the simulation results by both internal and external measures. The field experiment would reflect the contextual information embedded in the neighborhood residents' cognitive and behavioral performance and the effectiveness of the neighborhood IT communications infrastructures.

This analysis contributes to the integration of technical and social components of the tsunami early warning system. The "system" is often seen as a network of technical devices rather than a system that depends greatly on human capacities and skills and a common understanding of what to do and how to react. The disaster preparedness approach for a community evacuation is still limited to early warning infrastructure, which relies on using high-tech devices for tsunami detection in the open sea. Often neglected and ignored, however, is the human cognitive factor, which is critical to strengthen coordination and cooperation among multiple, diverse respondents. The neighborhood electronic communication system outlined in this book could be extended to other Indonesian cities exposed to coastal hazards to increase their resilience to tsunami hazards.

References

Ai, F., L. K. Comfort, Y. Dong, and T. Znati. 2016. "A dynamic decision support system based on geographical information and mobile social networks: A model for tsunami risk mitigation in Padang, Indonesia." *Safety Science* 90: 62–74.

Andersen, T., J. Carstensen, E. Hernandez-Garcia, and C. M. Duarte. 2009. "Ecological thresholds and regime shifts: Approaches to identification." *Trends in Ecology and Evolution* 24, no. 1: 49–57.

AnyLogic. 2018. "What is system dynamics modeling?" https://www.anylogic.com/use-of-simulation/system-dynamics/.

Bellifemine, F., F. Bergenti, G. Caire, and A. Poggi. 2005. "JADE—a java agent development framework." In *Multi-Agent Programming, Multiagent Systems, Artificial Societies, and Simulated Organizations*, vol. 15, ed. R. H. Bordini, M. Dastani, J. Dix, and A. El Fallah Seghrouchni, 125–47. Boston: Springer.

Birkmann, J., N. Setiadi, and N. I. K. L. A. S. Gebert. 2008. "Socio-economic vulnerability assessment at the local level in context of tsunami early warning and evacuation planning in the city of Padang, West Sumatra." Conference paper presented at the International Conference on Tsunami Warning, Bali, Indonesia, November 12–14, 2008. https://www.researchgate.net/publication/230625596.

Borshchev, A., and A. Filippov. 2004. "From system dynamics and discrete event to practical agent based modeling: Reasons, techniques, tools." In *Proceedings of the 22nd International Conference of the System Dynamics Society*, July 25–29, 2004, Oxford, England. http://www2.econ.iastate.edu/tesfatsi/systemdyndiscreteeventabmcompared.borshchevfilippov04.pdf.

BPSKP (Badan Pusat Statistik Kota Padang). 2013. Padang Barat in Figures 2013, Badan Pusat Statistik Kota Padang. Retrieved from http://padangkota.bps.go.id.

Comfort, L. K. 1999. *Shared Risk: Complex Systems in Seismic Response*. Oxford, UK: Elsevier.

Comfort, Louise K. 2005. "Risk, security, and disaster management." *Annual Review of Political Science* 8: 335–56.

Comfort, L. K. 2007. "Crisis management in hindsight: Cognition, communication, coordination, and control." *Public Administration Review* 67: 189–97.

Comfort, L. K., N. Oh, G. Ertan, and S. Scheinert. 2010. "Designing adaptive systems for disaster mitigation and response." In *Designing Resilience: Preparing for Extreme Events*, ed. L. K. Comfort, A. Boin, and C. C. Demchak, 33–61. Pittsburgh: University of Pittsburgh Press.

Croft, D. G., S. P. Banbury, L. T. Butler, and D. C. Berry. 2016. "The role of awareness in situation awareness." In *A Cognitive Approach to Situation Awareness: Theory and Application*, ed. S. Tremblay and S. Banbury, 82–103. London: Routledge. First published in 2004 by Ashgate by S. Tremblay and S. Banbury, eds.

Endsley, M. R. 1995. "Toward a Theory of Situation Awareness in Dynamic Systems." *Human Factors Journal* 37, no. 1: 32–64.

Fracker, M. L. 1991. *Measures of situation awareness: Review and future directions.* Wright-Patterson Air Force Base, OH: Armstrong Laboratory.

Fransen, J., P. A. Kirschner, and G. Erkens. 2011. "Mediating team effectiveness in the context of collaborative learning: The importance of team and task awareness." *Computers in Human Behavior* 27, no. 3: 1103–13.

Galy, E., and C. Mélan. 2015. "Effects of cognitive appraisal and mental workload factors on performance in an arithmetic task." *Applied Psychophysiology and Biofeedback* 40, no. 4: 313–25.

Gary, M. S., M. Kunc, J. D. Morecroft, and S. F. Rockart. 2008. "System dynamics and strategy." *System Dynamics Review: Journal of the System Dynamics Society* 24, no. 4: 407–29.

Hale, J. E., R. E. Dulek, and D. P. Hale. 2005. "Crisis response communication challenges: Building theory from qualitative data." *Journal of Business Communication*, 42, no. 2: 112–34.

Hoppe, M., and H. S. Mahadiko. 2010. "30 minutes in Padang—lessons for tsunami early warning and preparedness from the earthquake on 30 September 2009." GTZ-GITEWS project publication. www.gitews.org/tsunami-kit.

Jackson, S. A., S. Kleitman, P. Howie, and L. Stankov. 2016. "Cognitive abilities, monitoring confidence, and control thresholds explain individual differences in heuristics and biases." *Frontiers in Psychology* 7: 1559.

Jax, K. 2014. "Thresholds, tipping points and limits." *Open-NESS Reference Book. European Commission Seventh Framework Programme (FP7) Grant Agreement No.* 308428. www.openness-project.eu /library/reference-book/sp-thresholds.

Klein, G. A. 1989. "Recognition-primed decisions." In *Advances in Man-Machine Systems Research*, ed. W. B. Rouse, 47–92. Greenwich, CT: JAI.

Klimoski, R., and S. Mohammed. 1994. "Team mental model: Construct or metaphor?" *Journal of Management* 20, no. 2: 403–37.

Kostoulas, D., R. Aldunate, F. P. Mora, and S. Lakhera. 2008. "A nature-inspired decentralized trust model to reduce information unreliability in complex disaster relief operations." *Advanced Engineering Informatics* 22, no. 1: 45–58.

Lyons, R. F., K. D. Mickelson, M. J. Sullivan, and J. C. Coyne. 1998. "Coping as a communal process." *Journal of Social and Personal Relationships* 15, no. 5: 579–605.

Lloyd, M., and A. Alston. 2003. "Shared awareness and agile mission groups." 8th International Command and Control Research and Technology Symposium (ICCRTS). Washington, DC: National Defense University, Command and Control Research Program.

Mayer, Richard E., and Logan Fiorella. 2021. *Cambridge Handbook of Multimedia Learning*, 3rd ed. Cambridge, UK: Cambridge University Press.

Mileti, D. 1999. *Disasters by Design: A Reassessment of Natural Hazards in the United States.* Joseph Henry Press.

Nofi, A. A. 2000. *Defining and Measuring Shared Situational Awareness* (No. CRM-D0002895. A1). Alexandria, VA: Center for Naval Analyses.

North, D. 2005. *Understanding the Process of Economic Change.* Princeton, NJ: Princeton University Press.

Nowell, B., and T. Steelman. 2015. "Communication under fire: The role of embeddedness in the emergence and efficacy of disaster response communication networks." *Journal of Public Administration Research and Theory* 25, no. 3: 929–52.

Nowell, B., T. Steelman, A-L. Velez, and A. Yang. 2018. "The Structure of Effective Governance of Disaster Response Networks: Insights from the Field." *American Review of Public Administration* 48, no. 7: 699–715.

Ostrom, E. 2005. "Self-governance and forest resources." *Terracotta reader: A market approach to the environment* 12.

Rahayu, H. 2016. Email message to first author, February 19.

Riad, J. K., F. H. Norris, and R. B. Ruback. 1999. "Predicting evacuation in two major disasters: Risk perception, social influence, and access to resources 1." *Journal of Applied Social Psychology* 29, no. 5: 918–34.

Riadi, Y., and J. Griffiths. 2018. "Indonesia tsunami: Grim search for survivors continues as death toll reaches 430." *CNN*, December 26, 2018. https://www.cnn.com/2018/12/25/asia/indonesia -tsunami-intl/index.html.

Setiadi, N., and J. Birkman. 2010. "Socio-economic vulnerability indicators." In *Rahmen Last-Mile— Evacuation Projekts*. Arbeitspaket 1000. Bonn, Germany: United Nations University, Institute for Environment and Human Security.

Spahn, H., M. Hoppe, H. D. Vidiarina, and B. Usdianto. 2010. "Experience from three years of local capacity development for tsunami early warning in Indonesia: Challenges, lessons and the way ahead." *Natural Hazards and Earth System Sciences* 10, no. 7: 1411–29.

Simon, H. A. 1996. *The Sciences of the Artificial*, 3rd ed. Cambridge, MA: MIT Press.

Taubenböck, H., N. Goseberg, G. Lämmel, N. Setiadi, T. Schlurmann, K. Nagel, F. Siegert, et al. 2013. "Risk reduction at the 'Last-Mile': An attempt to turn science into action by the example of Padang, Indonesia." *Natural Hazards* 65, no. 1: 915–45.

Van Kleef, G. A., and A. H. Fischer. 2016. "Emotional collectives: How groups shape emotions and emotions shape groups." *Cognition and Emotion* 30, no. 1: 3–19.

Wellens, A. R. 1993. "Group situation awareness and distributed decision making: From military to civilian applications." In *Individual and Group Decision Making: Current Issues*, ed. N. John Castellan, 267–91. New York: Psychology Press.

Chapter 7

Wireless Networks for Disaster-Degraded Contexts: Tsunami Evacuation in Padang, Indonesia

Fuli Ai, X. Xerandy, Echhit Joshi, Taieb Znati, and Febrin Anas Ismail

The continuing threat to communities in coastal regions from near-field tsunamis is a major policy concern for Indonesia, with many island coastlines very close to tsunami sources. Building communities that are resilient to hazards is a long-term project, requiring the commitment and engagement of the whole nation. This task necessarily includes scientists, but also government agencies, private and nonprofit organizations, and informal community groups. It requires a sociotechnical framework composed of technical, organizational, cultural, and socioeconomic subsystems to manage resilience to hazards. Importantly, it involves the recognition of risk by individuals, and collectively, by the community, to form the basis for collective action to reduce risk and enhance safety for the community. Building resilience means transforming societal understanding of risk to support systematic monitoring of risk conditions, dynamic data analysis, decision making, and communication needed to guide collective action in communities at risk.

Information Technologies Offer Decision Support

Information technologies offer technical means of providing decision support, coordinating approaches to resolve conflict, and disseminating timely policies. Implementing these strategies in practice is not trivial. Recent studies illustrate the difficulty of creating awareness of shared risk. For instance, in 2007, an earthquake and tsunami occurred south of South Pagai, one of the Mentawai Islands off Sumatra, Indonesia. No one was killed, as the population self-evacuated in the coastal area. Then, in

167

2010, the Mentawai earthquake and tsunami killed more than 400 people in the same region. The Tsunami Research Center at the University of Southern California (2010) compared these two disasters. In the 2010 event, local residents in the Pagai Islands had been told to self-evacuate only if the shaking was stronger than the 2007 earthquake. The 2010 Mentawai earthquake generated a slow, rolling motion but no sharp jolt, so some people did not even get out of bed. Also, there was no warning disseminated to the Mentawai Islands. Residents in Padang saw an announcement on television but did not evacuate because the announcer only warned of the potential of a tsunami. Most residents in Sikakap, a subdistrict on the island of North Pagai in the Mentawai Islands Regency off the coast of West Sumatra, did not see the announcement, although a few were alerted by relatives who telephoned them from elsewhere in Sumatra. In most coastal villages in the Pagai Islands and Sipura, the capital of the Mentawai Islands Regency, there is no electricity or telephone communications, so residents could not receive the televised alert. Residents in some villages had taken part in tsunami drills or public education, but not all villages shared an awareness of tsunami threats. Even residents who were well-informed regarding tsunami risk and self-evacuation strategies failed to take action because the harbinger signs of the impeding catastrophe were not felt or observed in a manner that would have triggered a "flee or die" response.

Individual action is subject to limited individual cognitive capacity, which relies on both short-term memory and long-term memory (Rittel and Weber 1973; Simon 1996). Updating these concepts in problem-solving mode, Kahneman (2012) refers to problem solving that uses immediate, short-term memory as system 1 but accesses a larger store of information through long-term memory as system 2. System 1 is fast, but often error prone. System 2, in contrast, accesses a wider range of knowledge, but also takes more time for recall, association, and analysis. Using system 2, people can update information and correct mistakes, but under threat, the decision may come too late. In emergency situations, insufficient information and uncertain conditions increase an individual's cognitive load and mental stress, which may disrupt long-term memory and limit the cognitive process. Individuals must make decisions quickly within short-term memory, which may cause mistakes. The more useful the information received, the clearer is the emergency situation; mental pressure is reduced, and cognitive load is decreased. The individual then learns by using long-term memory to update decisions and correct mistakes, a function of system 2. The whole process forms the individual's situational awareness; increasing individual cognitive capacity will enhance informed, collective emergency action.

When individuals effectively connect through social networks, like Twitter, Facebook, and WhatsApp, with family members, neighbors, schoolmates, and work colleagues, they share positive emergency information through real-time communication. This exchange increases collective situational awareness, which increases capacity for collective action and community resilience. As explained in chapter 6, sharing timely critical information and resources during tsunami evacuation bolsters collective action among residents. A communications system designed for use in response to tsunami risk should be sufficiently resilient to support emergency communication, especially when normal modes of communication are not functioning, as in the breakdown of internet and cell phone networks. This chapter explicitly focuses on the design of the bowtie architecture identified in chapter 1 to enable residents of coastal communities to communicate under urgent conditions of near-field tsunami risk.

Conceptual Framework

The bowtie architecture (Csete and Doyle 2004) outlined in chapter 1 is designed to communicate critical information from multiple sources to create a common profile of risk for the whole community and communicate relevant information to different actors in a dynamic environment. The processes represented in this architecture include collecting and extracting information from multiple sources, integrating the extracted information into a common format, analyzing the information from the perspective of the whole community, and guiding organizational action in a regional information infrastructure. This model improves the capacity for communication and coordination among multiple organizations that confront a common threat (Comfort 2005).

Since Indonesia began to build its Indonesian Tsunami Early Warning System (InaTEWS) (BMKG 2012) following the 2004 Sumatran earthquake and tsunami, the national system, established in 2008, has been tested in actual events and continues to undergo evaluation and revision. Padang City, West Sumatra, developed a city contingency plan in 2013 to ensure correspondence at the local level with the national plan, but many critical aspects, including the confirmation of near-field tsunamis, rapid communication of threat, and instructions for evacuation among the population, are still lacking at the local level.

We propose to adapt and test the bowtie architecture to fit the working contingency plan for rapid communication of threat and evacuation support at a field study site in Padang. The communication infrastructure includes a Raspberry Pi network and satellite link to disseminate evacuation strategies and extend the communication network at the community level and a mobile, geographic information system (GIS)–based device application as social media. The infrastructure integrates information from community residents through search-and-exchange functions to build community cognition and collective awareness and then alerts all affected residents to risk when hazards occur. This process activates community resilience by enabling cooperation among community residents, facilitating coordination among multiple organizations, and mobilizing response to hazardous events as they occur.

This chapter proceeds as follows. First, we provide the introduction to the problem of collective communication under urgent stress, followed by an outline of the bowtie-based emergency communication infrastructure and its key properties for operating in disaster-degraded environments. We then present the design of the proposed infrastructure, focused on a Raspberry Pi–based wireless communication network for mobile devices, including a handover scheme, dynamic location tracking, mobility connection with emergency messaging, and dynamic leader election, and show how the proposed sociotechnical infrastructure would support community resilience through a set of simulated cases. We end with a set of conclusions and recommendations for further work.

Characteristics of a bowtie-based emergency communication infrastructure

As outlined in chapters 1 and 6, we use the bowtie model as the framework for the sociotechnical infrastructure to structure the information flow and provide decision support in risk conditions that require collective action. At the district and subdistrict levels, the lowest jurisdictional units in the administrative hierarchy, information is collected from different individuals and transmitted to a central processing unit in the municipal emergency operations center (EOC). These data are integrated, analyzed, and interpreted in the processing unit, and the output is disseminated to responsible leaders in the community to alert others and take collective action to reduce the threat. Their information and actions at the district level, in turn, are sent to the municipal/provincial/ national jurisdictional levels, and the integration, aggregation, and visualization functions are conducted for the municipal/provincial/national levels. The process is iterative and allows information from one level of operations to support adjustments at their specific operational context, which, in turn, changes the operating conditions at subsequent levels in the system. As in complex adaptive systems of systems (CASoS) (Glass et al. 2012), this infrastructure has features that include adaptation, social awareness, and interactivity. We consider the CASoS design as an optimization system and focus on the following features.

Reliability. The adaptive feature of the bowtie model strengthens the reliability of the technical communication infrastructure and ensures that there is at least one channel available to transfer timely, critical information to community residents. The central processing unit integrates, analyzes, and interprets information from multiple sources. If the resilient framework adheres to community norms and culture, it forms a reliable social network between the public and local government, linked through trusted community leaders that are elected by the community and certified by the local government (Ai et al. 2016). Elected leaders are key nodes in the sociotechnical network. They have the responsibility to maintain the connections and need to be available during an emergency. If the leader is not available or is out of the region, a dynamic leader election process is needed to avoid connection failure in the community network. The community prepares a list of candidate leaders prior to hazard occurrence, with backup leaders if needed, to activate an ordered leadership mechanism to support timely, collective action when hazards occur. Leaders with more experience and skills are placed at the top of the list and are more likely to be elected as the activity leader. Such a process facilitates a smooth flow of information among community residents.

Cost-effectiveness. We use cost-effective hardware such as single-board computers (SBCs) to create an emergency communication framework to disseminate information. To avoid extra investment, we also use an existing mobile device—a smartphone—as the communication medium for the public, and we develop cost-effective software for emergency communication. This communications framework allows "light-weight" messaging through multicast, not broadcast, transmission. During an emergency, the intense increase in information exchange will burden the communication infrastructure; all messages may be blocked, and none could be sent and received. Given these anticipated conditions, the extensive use of preloading and opportunistic loading of static information is critical. The static information, including critical facilities and landmarks, is cached in the local terminal device as preloaded information.

Dynamic information, including updated warnings and emergency messaging, is limited by packet size and sent through the infrastructure. Dynamic information will be integrated with preloaded static information using a message template. In nonemergency time, users update the static information. During emergency communications, users will send only dynamic critical information, such as ID and required parameters. All members share the same information codes to form a complete message. This simplified message template reduces the redundancy burden on the communication infrastructure and saves critical time during tsunami evacuation. The information will be sent to designated individuals, following the routing table, where only selected communication links and nodes in the network are used. This strategy reduces network traffic significantly. Various functionalities such as an initial warning with a default path to shelter, dynamic mobility tracking, real-time situation monitoring with detours, and medical and evacuation assistance will enable individuals to mobilize collective action and increase community resilience.

Timeliness. The bowtie framework supports two-way, real-time communication. From its top-down branch, local governments can disseminate disaster information and evacuation policies to community residents from the central unit in real time (Ai et al. 2016). This function reduces individual cognitive load during urgent events and increases individual situational awareness. With timely, valid information from the local government, residents can communicate with one another in their neighborhoods, form collective situational awareness about tsunami risk, and mobilize collective action through cooperation, coordination, and control (Comfort 2007).

From the bottom-up branch, residents in the community serve as environmental sensors and report the real-time situation to the local government. Direct observation of conditions in the neighborhoods reported by residents helps emergency managers collect timely information, analyze the changing situation, make decisions, and disseminate updated policies to the community. The shared emergency policies should be direct and clear; residents may not recognize the threat or have time to analyze the situation and make decisions. Most helpful are direct guidelines regarding what to do when and how, not why, to act (Mileti et al. 2002). For instance, the tsunami evacuation warning shows on the mobile device with a warning icon and vibration. This warning has a clear meaning for individuals to understand.

The GIS-based evacuation map provides an intuitive interface with familiar environmental information, such as buildings, road networks, and known facilities. It means "Follow the default evacuation route shown on the map to the allocated shelter now." The message is easy for the individuals to follow, as it is critical to increase collective evacuation time through decreasing cognition time. Also, using default buttons for instant communication saves time; a resident needs to push just one button to request emergency assistance, saving response time and reducing error.

Technical Design and Components

We designed a wireless communication network using SBCs for alternative access points to create an alternative channel for mobile devices to disseminate alert messages at the neighborhood level in an extreme event. The following section briefly describes an overview of a community emergency communications network, the technical components that are critical to a local neighborhood network, the design and operational functions of the local network using mobile phones, management of the mobility messaging functions, dynamic election of local leaders, and messaging functions for small groups. The overall technical goal is to use simple, low-cost components in a sophisticated networked design to achieve interactive communication among residents at the sub-subdistrict or neighborhood level of a community at risk.

Community emergency communications network

As the name implies, an SBC is a complete, functional computer system that is built onto a single, small, compact circuit board. With the advance of integrated circuit technology, this device can be built within a credit-card-size board, which also makes the cost of SBCs affordable, especially for a general-purpose computer. Although it is a functional computer system, an SBC has restrictions on CPU speed, memory, and storage space, so it is not suitable for tasks requiring high-performance computation. SBCs find their best fit in general applications, despite their limited resources for computing. SBC devices are widely used for simple application development, low-cost controllers, education, and emergency situations (Khelil et al. 2014; Hu and Huang 2015; Martínez et al. 2016).

Raspberry Pi is an SBC device that is widely known for such uses. This device was initially developed by the Raspberry Pi foundation, a United Kingdom–based charity, to promote teaching of basic computer science in schools. The foundation has released several versions of Raspberry Pi; the latest version is Raspberry Pi Zero 2 W, which was released in October 2021.

Local communications network

We explore the potential use of the Raspberry Pi for alternative communication access points during extreme events. We developed a network of Raspberry Pi devices to extend access to Wi-Fi coverage within a region via a USB dongle, a small tool that connects to a port on a computer or other device, to provide Wi-Fi signal. To implement a local communications network, the Raspberry Pi devices are deployed by attaching them to electrical poles along the street. Having the Raspberry Pi devices installed at higher points improves the hotspot effective coverage because it minimizes signal obstruction by large objects. This neighborhood electronic network can be connected to the local EOC through a satellite link.

We envisage the network as an alternative channel for the local EOC personnel to transmit timely alert messages at the community level in an extreme event, especially when the main communication infrastructure is damaged. This alternative channel would extend the bowtie architecture to the neighborhood level (Ferdoush and Li 2014). By connecting these devices in a multicast network in practice, network processing rates and energy consumption are significantly improved, and the network resource utilization can potentially reach its maximum instantaneously (Paramanathan et al. 2014). A more detailed description of developing a Raspberry Pi network for this purpose is given in chapter 4.

Relying on opportunistic communication and epidemic spreading of broadcast information, regular social media apps equipped with a disaster mode function would allow communication to resume instantaneously and help distressed people self-organize until regular communication networks are operational again (Hossmann et al. 2011). At the end-user side, mobile devices installed with an application exclusively built for this network can be used to gain access to the Raspberry Pi network. The interconnected Raspberry Pi devices create a Wi-Fi hotspot to which the mobile phones can connect. The Raspberry Pi network is linked to satellite communication, so it will provide a communication link between the local EOC and community residents when other means are not available, such as cell phones or internet. This network also provides horizontal communication within the community to coordinate collective evacuation.

We choose a user datagram protocol (UDP) to provide a communication link between the Raspberry Pi and mobile device, which is a simple, message-based, connectionless protocol, documented in RFC 768 (Postel 1980). It supports multicast data transmission, a single, lightweight datagram packet that can be automatically routed without duplication to a very large numbers of subscribers. This protocol matches the emergency communication demand to disseminate disaster alerts to targeted communities at risk.

We then present the design of the proposed infrastructure, focused on a Raspberry Pi–based wireless communication network for mobile devices, including a handover scheme, dynamic location tracking, mobility connection with emergency messaging, and dynamic leader election, and show how the proposed sociotechnical infrastructure would support community resilience through a set of simulated cases.

Mobility connection management

A main technical challenge in designing a Wi-Fi based network is reliable connection among the mobile devices during movement. We designed a dynamic location tracking scheme and handover scheme in the Raspberry Pi network to achieve real-time location tracking of mobile devices.

Mobility-based tracking. The challenging aspect in managing a local network is incorporating the user's mobility. This aspect involves both the Raspberry Pi network and mobile devices. From the network perspective, the Raspberry Pi should track a mobile device's location. In this network, a mobile device's location is identified by the Raspberry Pi to which the mobile device is currently connecting. To perform this tracking scheme, a Raspberry Pi is configured as a *Home Agent*. The Raspberry Pi tracking scheme is similar to the Home Location Register (a database used to store identity and location information that enables charging and routing) in a cellular system or base station (Iyer and Iyer 2009; Nakauchi and

Shoji 2016). Consequently, the mobile device needs to update its location when the user moves; its address is maintained in a *Home Agent* through a *Current Agent* (also called care of address, or CoA).

Mobile devices stay in their original Wi-Fi networks to prevent devices from consuming too much energy in interface switching (Yang et al. 2009). We present a Wi-Fi–based infrastructure wireless mesh network (Boukerche and Zhang 2008; Boukerche et al. 2009; AlShamaa et al. 2018) at the sub-subdistrict level of the municipal jurisdiction as the basic operational unit. This technical infrastructure could extend to subdistrict and district levels of jurisdictional administration and support real-time communications, moving across geographic districts with a low ratio of packets lost to packets sent. This infrastructure adaptively constructs a GIS-based mobility map of the environment (Zhou, Tian et al. 2014; Zhou, Xu et al. 2015) using a shortest path routing algorithm to minimize packet forwarding latency. Both intradomain and interdomain handoffs provide seamless roaming. Further description on the tracking scheme from the Raspberry Pi network perspective is described in chapter 4.

The mobile device needs to monitor its location regularly. The mobile device gains access to the local network by connecting to the Wi-Fi hotspot created by the Raspberry Pi. Every hotspot in the Raspberry Pi network has been assigned a single service set identifier (SSID), a thirty-two-character unique identifier attached to the header of packets sent over a wireless local area network that acts as a password when a mobile device tries to connect to the basic service set (BSS). We use the single-SSID scheme to speed up connection time between Raspberry Pi and mobile devices. Consequently, in this single-SSID scheme, a mobile device distinguishes the Raspberry Pi through the device's medium access control address or basic service set identifier (BSSID), a forty-eight-bit identification tag used to identify a particular BSS within an area, at the link layer and Raspberry Pi ID at the application layer.

Connection handover scheme. When a mobile device is moving from one Pi to another, the Wi-Fi connection will go through a connect-disconnect-connect process. Usually, the Wi-Fi hardware in the mobile device will scan the available Wi-Fi signals around it, list them based on signal strength, and then connect to the strongest signal. The mobile device essentially checks its Wi-Fi connection states as managed by the mobile device kernel. The kernel will notify the application if the connection status has changed from either connected or disconnected. We cannot control how fast the kernel will respond upon the change, so the time for response heavily depends on the mobile device connection's processing time.

To avoid unnecessary scanning and connection processes, we devised a handover mechanism based on codes that are available on request. All Raspberry Pis are configured with the same SSID ("RasPdg") and different BSSID. All mobile devices are assigned the same static internet protocol (IP) address, for instance, "192.168.64.112," for UDP and use the group IP to multicast. That means all Pis in the mesh network are considered to be the same Pi for the mobile devices. When a mobile device moves in an overlapping Pi network or when the Pis are nonoverlapping but within a short distance, the Wi-Fi signals of the previous Pi and current Pi will not change a lot. The mobile device will function as if it is still in the same Pi coverage and will keep the connection seamlessly and with low latency. Omitting the association process, the mobile device will use the whole connection stage for messaging. As the status continues, the mobile device will evaluate if there are any changes on the BSSID. If a change on the BSSID is detected, the mobile device initiates its tracking scheme by performing the three-way handshake association process to the Raspberry Pi and updates its Home Agent.

Reliable connection scheme. As shown in figure 7-1a, when a mobile device moves towards a Raspberry Pi, the Wi-Fi signal increases. In the center t_0, the Wi-Fi signal is strongest; as the mobile device moves away from the Pi, the Wi-Fi signal becomes weaker and the packet loss rate increases. In t_1, as the packet loss rate increases, the signal becomes too weak to receive

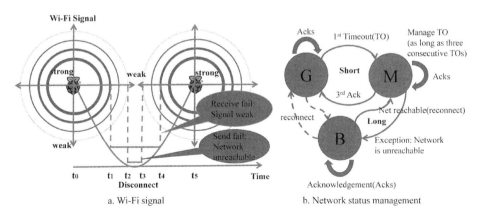

FIGURE 7-1. Mobile device moving and connection process: (a) Wi-Fi signal; (b) network status management.

or send information or, at least, disconnects with the Pi. In t_2, when the mobile device moves toward another Pi, the signal increases again. The signal curve in one Pi is like a sine wave. The periods between t_0 and t_1, t_1 and t_2, t_2 and t_3 in figure 7-1b are called *Good Channel* (*G*), *Maybe Bad Channel* (*M*), and *Bad Channel* (*B*), respectively.

We use a mechanism to enhance the association process to increase stability and timeliness of the tracking and messaging process during the Maybe Bad and Bad Channel periods. Usually, access points (APs) bridge traffic between a mobile device and other devices on the network. Before a mobile device can send traffic through an AP, it must be in the appropriate connection state. The three 802.11 connection states are (1) not authenticated or associated, (2) authenticated but not yet associated, and (3) authenticated and associated. A mobile device must be in an authenticated and associated state before bridging will occur. The mobile device and AP will exchange a series of 802.11 management frames to get to an authenticated and associated state, a three-way handshake as shown in figure 7-2a.

UDP is unreliable. When a UDP message is sent between Raspberry Pi and a mobile device, it cannot be known if it will reach its destination. It could get lost along the way; there is no concept of acknowledgment, retransmission, or timeout. In the Raspberry Pi network, we use a three-way handshake method to create the connection between a mobile device

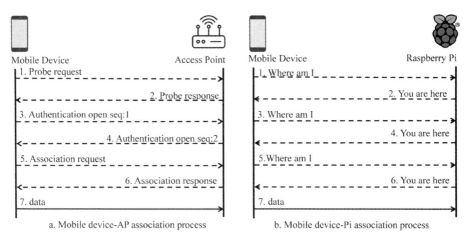

FIGURE 7-2. Wi-Fi association process: (a) Mobile device–access point association process; (b) mobile device–Pi association process.

and Raspberry Pi (figure 7-2b). Shown in figure 7-3, as default, the mobile device will ask for the current Pi with an optimized short timeout, 2-second interval. If it receives a response three times from the current Pi, it is in the Pi's coverage with a Good Channel, the periods between t_0 and t_1, and it can send/receive emergency messages. If the mobile device sends a message but does not receive all acknowledgments with an optimized timeout, it is moved from t_1 to t_2 in the Maybe Bad Channel. It will keep increasing the timeout to 4, 8, 10 seconds—10 seconds being the maximum timeout—to give enough time to receive acknowledgments with a weak signal. Further explanation regarding the technical details of the neighborhood electronic network are provided in an appendix to chapter 7, posted on the online directory established for this book, https://islandpress.org/tsunami. Additional information regarding the UDP protocol is provided in the online appendix, table 2a. If the timeout is kept at maximum value, the mobile device could not reach the wireless connection or disconnects with the Pi and thus moves to the Bad Channel. From t_2 to t_3, the mobile device could not send or receive messages. If it receives acknowledgments in a maximum timeout, the signal becomes stronger, and it will decrease the timeout by 8, 4, 2 seconds until it reaches the Good Channel. Further details to explain this connection process are given in the online appendix, note 2 and table 2, https://islandpress.org/tsunami.

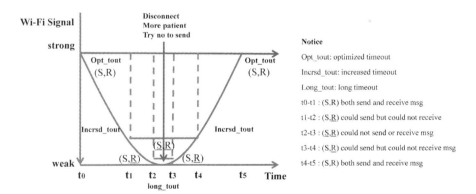

FIGURE 7-3. Connection timeout control process. (Note: t_0–t_1: both send and receive message; t_1–t_2: could send but could not receive; t_2–t_3: could not send or receive message; t_3–t_4: could send but could not receive message; t_4–t_5: both send and receive message.)

Mobility-based messaging

During the evacuation, when communication occurs between two moving users based on the locations of the two users obtained from the tracking process, the communication process allows four options for connection:

> Option 1. If the sender and the receiver are in the same *Current Pi*, they can exchange messages with each other through the same Pi directly (figure 7-4a).
>
> Option 2. If the receiver is still in its *Home Pi*, the *Home Pi* will forward the message to that receiver directly (figure 7-4b).
>
> Option 3. If both the sender and the receiver are moving and staying in their current Pis, the receiver's *Home Agent* forwards the information to the receiver's *Current Pi,* including the sender's current Pi ID. At the same time, the receiver's *Home Agent* sends the receiver's current Pi ID to the sender's *Current Pi*, and then the sender and receiver both know each other's current location. When the sender receives the information forwarded from the receiver's *Home Agent Pi*, the message is sent to the receiver's *Current Pi*. When the receiver receives the message, the receiver replies directly to the sender's *Current Pi* (figure 7-4c).
>
> Option 4. If both the sender and the receiver are moving, they do not exchange messages through *Current Pi* directly. Instead, they will send message to each other's *Home Agent Pi*, and the *Home Agent Pi* will forward the message to their *Current Pis* (figure 7-4d).

During the messaging process, UDP packets are sent individually and are checked for integrity only if they arrive. Packets have definite boundaries that are honored upon receipt, meaning that a read operation at the receiver socket will yield an entire message as it was originally sent. A normal complete packet is formatted as in table 7-1. The receiver needs to extract the receiver ID part and check if the packet is sent to them; if yes, they will further process the packet. UDP packets do not arrive in order; if large or group messages are sent to the same receiver, the order in which messages arrive cannot be predicted. The packet has a length limitation, as shown in table 7-1, of 50 bytes. If a packet is 100 bytes, it will be divided into two packets with sequence ID #1 and #2 and will be sent out separately. When the receiver receives #1(#2), it will wait until #2(#1) arrives to form the complete packet and then process both packets together. If the mobile device does not receive the second packet before a timeout, it will drop the first packet.

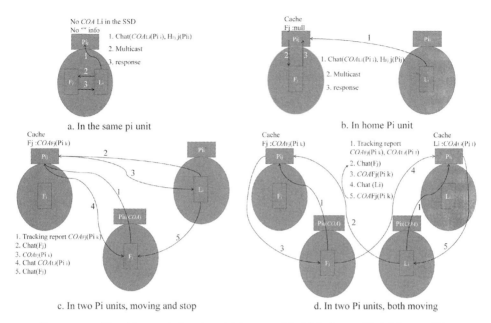

a. In the same pi unit

b. In home Pi unit

c. In two Pi units, moving and stop

d. In two Pi units, both moving

FIGURE 7-4. Tracking and chatting: (a) in same Pi; (b) in home Pi; (c) in two Pis, moving and stop; (d) in two Pis, both moving.

During an emergency evacuation, there will be many-to-one and even many-to-many messages sent in the same time period. For instance, many followers send requests for emergency assistance to the same community leader, but the UDP does not have a congestion control scheme; in other words, the UDP itself does not avoid congestion. Congestion control measures must be implemented at the application level. Technically, information packets broadcast by mobile devices will be subsequently repeated by other receiving mobile devices. To prevent collision, each mobile device assigns a random probability to determine the timeout before initiating transmission. The mobile device should defer its transmission immediately once it hears other units already transmitting the information. In a longer run, this mechanism will enable rapid dissemination of information among neighboring wireless devices (Biswas and Morris 2004). In this design, the mobile device application uses the producer-consumer-queue multiple threads method (Arpaci-Dusseau and Arpaci-Dusseau 2018). Consider the connection scheme discussed above, with the producer as a receiver set up for timeouts during an incoming packet listening process to avoid congestion. This scheme puts the received packet into different queues based on type of reception

TABLE 7-1. Packet format structure

Component	Length (bytes)
Length	4
Type	4
Sender ID	4
Receiver ID	4
Sequence ID	4
Time stamp	4
Payload	16
Option	10

shown in table 7-1; one consumer works with one queue to further process the packet.

Dynamic leader election

To maximize rapid dissemination of emergency information to community members and to enhance collective cognition of risk, we use the current community emergency leader-follower social network and introduce the leader election mechanism. This mechanism is designed to elect the activity leader of the community at any time. A given leader may not be in the community region all the time, so when the selected leader is out of the region for any reason, like travel, there must be at least one candidate leader available to be the alternative leader to communicate emergency information to the community.

As default, each community has a candidate leader list ordered by the leaders' tsunami experience scores, registered with the local government agencies (Ai et al. 2016). Some candidates are leaders of local nongovernmental organizations. Leaders in one subdistrict share one key value dataset, and each level has an allocated number of leaders. When a leader registers and enters their name, the central coordinator in the municipal EOC will allocate a leader ID for that person and put <leader level, leader id, region id,...> as the value and save in the key value store (KVS) in the municipal EOC (Han et al. 2011).

When the emergency information is sent from the municipal EOC to a tsunami-affected community, the Raspberry Pi network will trigger the activity leader election process first. All candidate leaders in the region will receive the election message; each candidate leader will wait for a random time based on their leader score. If a timeout occurs and they do not receive a response from candidates higher on this list in their current Pi, they will send a response, "I want to be the leader." Each Pi will wait for a fixed time for responders from its own Wi-Fi region before electing leaders from more distant Pis. Then it will elect the first responder on the list with highest score as the activity leader, L1. If the default leader L0 is on the list, the leader mechanism will elect L0 as the activity leader and then inform all the members in the community through multicast. See further explanation on the leader election process in the online appendix, note 3 and figure 3, https://islandpress.org/tsunami.

The election process will occur many times during an evacuation, if the emergency information does not reach the activity leader, until all the community members reach safety in allocated shelters or until the tsunami warning is expressly ended.

Emergency messaging. As described in chapter 2, emergency messaging uses a default button and preloading message template. The local community resilience framework, Raspberry Pi, and the mobile device application share a message template and only send critical dynamic information and required parameters to reduce the burden on communication infrastructure, the Raspberry Pi local network, to avoid interference and save critical time. The default message button interface of the mobile device application is shown in the online appendix for chapter 7, figures 1 and 2, https://islandpress.org/tsunami. These are first-level class buttons—take action, evacuation assistance, medical assistance, and urgent supplies—shown in the bottom navigation menu in the mobile device application interface.

Leader messaging. The leader cannot communicate (chat) with all the followers at the same time. We use a messaging scheme to manage the emergency messaging process between the leader and followers to avoid message congestion. The leader uses one queue to hold all the received chat messages for each first-level communication and divides the messages into three types: new chatting request, continuous chatting, and closed chatting, shown in the online appendix, figure 4, https://islandpress.org/tsunami. If the leader establishes a new chat or response, that person needs to schedule one chat first, extract the messages from the queue, and process them one by one. In this design, one leader could chat with four followers at the same time. See the online appendix, figure 5 and tables 3 through 8. See further explanation on the leader chat function in the online appendix, note 4, https://islandpress.org/tsunami.

Follower messaging. When a follower pushes the chat button, that person needs to first establish a chat session with the leader. A unique session ID is dynamically created and links to the follower (online appendix, figure 6a). If that person has established the chat session, the chat interface will show. They could chat with the leader many times when the session is not closed and the connection is still stable (online appendix, figure 6b). Or they could close the chat session (online appendix, figure 6c) so that other followers have a chance to chat with the leader. See further explanation on the follower chat function in the online appendix, note 5 and tables 9 and 10, https://islandpress.org/tsunami. These procedures seek to balance access to communication for followers with time available for community leaders.

Session management. The leader needs to manage the chat session, which includes four options, listed in the online appendix, note 6. This process is especially important under urgent conditions when time is critical and communication is central to safe evacuation.

> Option 1: *Accept session.* When the active queue is not full, the leader allocates a session for the follower's request, adds the follower session ID to the active queue, and sends a message to the follower.
>
> Option 2: *Active queue is full, hold queue is not full.* Add request to hold queue if queue is not full; else, if hold queue is full, reject. Move requests from the hold queue to the active queue when the leader processes a request from the active queue.
>
> Option 3: *Follower receives no reply.* Show timeout and send request again.
>
> Option 4: *Eject session.* If active queue is full and new request has high emergency level, eject the request that has the longest idle time in the active queue; do not send notice to the ejected one. When the ejected one sends message, if the active queue is not full, create a new session for them. If not, eject the chatting request.

Small group messaging

In some special small groups, members need communication during emergency situations. We consider here family, emergency, and dynamic groups.

Family group. All family members must register first to have a unique user ID. When the family's leader registers, that person will first search for the family group ID in the KVS based on last name and address. If the family group ID exists, the leader will join the group and update it as the family leader and inform existing members. If not, the leader will create a new family group ID and join the group as leader. When a family member registers, that member will first search for the family group ID in the KVS based on last name and address. If it exists, the family member joins the group and informs all other family members. If not, the family member will create a family group as the temporary leader, join the group, and update local family contacts when information is received. The family members could chat with targeted member(s), and all family members use default buttons during the emergency, including "Where are you?," "I am here," "Are you OK?," "I am OK." The interface is shown in figure 7-5. The codes are shown in the online appendix, table 13.

Emergency group. Shown in figure 7-6, the family leader registers, creates an emergency group with last name and address, fills in emergency contact form with last name and address, searches and gets emergency contact IDs, saves the form for the local EOC and mobile device, and informs emergency contacts. Family members register as spouse, child1, child2, and so on. During a disaster, the leader can send stress messages that demand immediate action(e.g., house on fire) to targeted member(s) or all emergency members. Because members in the emergency contact list may not know one another, stress messages should include the member

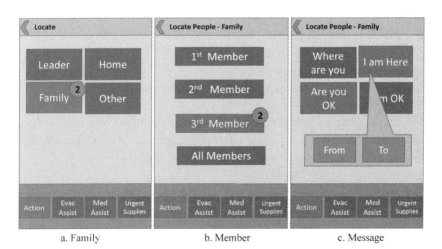

FIGURE 7-5. Family group messaging interface: (a) family; (b) member; (c) message.

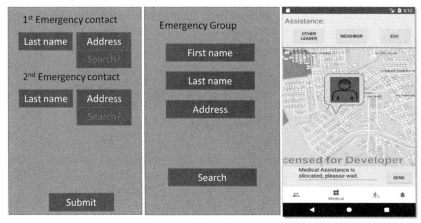

a. Family leader: Registration   b. Family member: Search group   c. Dynamic group

FIGURE 7-6. Emergency group messaging interface: (a) family leader, registration; (b) family member, search group; (c) dynamic group.

IDs. The emergency assistance process is shown in the online appendix, table 15.

Dynamic group. If a community leader wants to find help, that leader co-ordinates with other candidate leaders in the same community, leaders in neighboring communities, or the local EOC. If is necessary to create a dynamic emergency group in which there is no group leader, the closest responder will become the leader, and the logic is similar to the emergency group. The interface is shown in figure 7-6c.

Study Experiments and Results

We designed experiments in actual outdoor environments to test the features of the proposed communication infrastructure and to answer the following questions:

Reliability: What distance between Raspberry Pi and mobile device will maintain reliable connections?
Cost-effectiveness: How many phones could one Raspberry Pi support for a designated time with UDP socket communication?
Timeliness: What time is required for this infrastructure to support a real emergency evacuation?

Experiment 1: Reliable connections

Wi-Fi routers operating on the traditional 2.4 GHz band reach up to 46 meters indoors and 92 meters outdoors. Newer 802.11n and 802.11ac routers that operate on both 2.4 GHz and 5 GHz bands vary similarly in reach. We choose the EDIMAX-150Mbps Wireless IEEE802.11b/g/n nano USB adapter, which complies with wireless 802.11b/g/n standards with a data rate up to 150 Mbps and supports smart transmit power control and auto-idle state adjustment. The power consumption can be reduced from 20 to 50 percent, and the power consumption reduction increases wireless coverage three times further.

The easiest and most consistent way to express Wi-Fi signal strength is with dBm, which stands for decibels relative to a milliwatt (mW), usually −30 to −100; −30 is a stronger signal than −100. Note that dBm does not scale in a linear fashion but instead is a logarithmic measure, which means that signal strength changes are not smooth and gradual. An example of the logarithmic change in signal strength and suggested rates of change is shown in the online appendix, table 16.

Scenario. We set up one Raspberry Pi device with an EDIMAX Wireless USB adapter, attached to an electrical pole 2 meters above the ground in an open area, to provide a local Wi-Fi signal without environmental noise and one smartphone brand Samsung Galaxy J1 2016 that runs a UDP socket application to communicate with the Pi. The smartphone sends small data packets (less than 50 bytes), receives the acknowledgment with the same message ID, and logs the round-trip time (RTT) and the Wi-Fi signal strength. The Pi sends an acknowledgment immediately when it receives the request. The phone moves away from the Pi step by step, to a distance from 10 to 100 meters at 10-meter intervals.

Results. Based on the suggested Wi-Fi signal strength shown in the online appendix, table 16, −70 dBm is the minimum signal strength for reliable packet delivery, −80 dBm is the minimum signal strength for basic connectivity, and packet delivery may be unreliable. As shown in table 7-2 and the online appendix, figure 10b, at about 30 and 50 meters the Wi-Fi signal strengths are approximately −70 dBm and −80 dBm, separately. In the online appendix, figure 10a, the RTT plus standard deviation (SD) increases from 50 meters. In a real situation, considering the possibilities of potential environmental noise and many phones communicating with one Pi at the same time, the smartphone should stay less than 30 meters

TABLE 7-2. Communication distance, round-trip time (RTT), and signal strength

Distance (meters)	Mean RTT (SD) (milliseconds)	Signal Strength (SD) (dBm)
10	10.161 (11.538)	−65.023 (4.727)
20	11.173 (16.765)	−69.021 (4.000)
30	11.177 (20.318)	−71.204 (4.398)
40	11.242 (15.391)	−72.621 (3.878)
50	*15.437 (44.231)*	*−78.754 (2.151)*
60	13.708 (44.659)	−79.034 (2.297)
70	18.823 (107.848)	−84.037 (3.252)
80	13.867 (42.175)	−79.093 (5.315)
90	12.814 (26.468)	−79.030 (3.447)
100	13.038 (34.391)	−77.143 (4.876)

from the Pi to keep a reliable Wi-Fi signal for communication or no more than 50 meters for emergency communication, which is the minimum signal strength for basic connectivity.

Experiment 2: Cost-effective messaging

In transmission control protocol (TCP), a connection-oriented, reliable, byte stream–based transport layer communication protocol in internet-based networks, many individual wireless routers and other access points support up to approximately 250 connected devices. Routers can accommodate a small number of wired Ethernet clients with the rest connected over a wireless network, usually between one and four clients. The speed rating of access points represents the maximum theoretical network bandwidth they can support. A Wi-Fi router rated at 300 Mbps with 100 devices connected, for example, can only offer 3 Mbps to each of them on average ($300/100 = 3$).

UDP does not maintain device connection; the device connection capacity of the Raspberry Pi (access point) is hard to test, especially in multicast and broadcast modes. Our infrastructure is only used in an emergency; a single packet is small, less than 50 bytes, but when many mobile devices try to maintain the mobility connection with the Raspberry Pi network and use the scheme proposed for mobility connection management and messaging during the evacuation, many packets are sent in multicast mode in the network. This pattern could lead to an information explosion, which would affect the mobility connection and communication.

Scenario. Considering space limitations, if a Raspberry Pi is attached to an electrical pole along the road, as the test results show above in experiment 1, the stable Wi-Fi signal coverage is a 30-meter radius half circle, $3.14 \times 30^2/2 = 1,413$ m^2. During the evacuation, each person needs space to move; assuming that the minimum space per person is a 1-meter-radius circle, $3.14 \times 1^2 = 3.14$ m^2, the maximum number of mobile devices one Pi could support is around 450 (1,413/3.14). Considering communication technology, the following parameters are set as technology limitations to test the Raspberry Pi connection capacity:

$$\text{Reliability} > 90\%$$
$$\text{Timeliness} < 1 \text{ second}$$

The maximum number of mobile devices that one Pi could support is the smaller value between space and technology limitation.

We designed simulation scenarios that use both personal computers (PC) and smartphones in the experiment. Details are shown in table 7-3. Codes are shown in the online appendix, table 14.

Results. In PC-Pi case analysis results, shown in table 7-4 and the online appendix, figure 11a, the RTT keep increasing with the increase in the number of simulated smartphones. Shown in the online appendix, figure 11b, from 96 to 112, the packet success rate drops from 93 to 58 percent sharply, so the number of phones that one Pi could handle is between 90 and 100.

In Phone-Pi case, we assume that the ideal maximum number of phones is 100 and that each phone runs 1, 3, 5, 7, 9 threads separately. The results are shown in table 7-5 and online appendix, figure 12. When

TABLE 7-3. Capacity experimental scenarios

PC-Pi Case	Phone-Pi Case
Based on the results of section Experiment 1: reliable connections, use 2 PCs to run *N* threads. Each PC creates 1, 2, 4, 8, 16, 32, 36, 48, 56, 64 threads to simulate 2, 4, 8, 16, 32, 64, 72, 96, 112, 128 smartphones to communicate with one Pi. Each thread sends a packet and waits to receive acknowledgment.	Based on the results of the PC-Pi case, assume the ideal maximum number of phones is *N*. Use 10 mobile phones to test loads of 10%, 30%, 50%, 70%, and 90%. Each phone runs $N \times 1\%$, $N \times 3\%$, $N \times 5\%$, $N \times 7\%$, $N \times 9\%$ threads in each load. Each thread sends a burst size of message (three messages), then sleeps for a random time (maximum 10 seconds), then loops back.

TABLE 7-4. RTT and packet success rate in PC-Pi case

Number of Phones	RTT (SD)	Packet Success Rate (%)
2	6.714 (10.513)	99.689
4	9.838 (14.739)	99.526
8	13.738 (19.006)	99.377
16	22.120 (28.149)	99.156
32	44.197 (59.238)	98.741
64	65.192 (101.695)	98.123
72	74.425 (131.319)	97.941
96	75.301 (148.847)	92.996
112	130.653 (209.005)	58.168
128	129.339 (231.751)	8.878

the load is less than 70 percent, the sum of RTT and SD is less than 400 milliseconds, and the packet success rate is above 99 percent. Thus, in a real situation, one Pi could support at least 70 people smoothly. When the load increased to 90 percent, the summed time (122.073 + 960.857) is more than 1 second, longer than the experimental technology limitation. Considering both the space and technology limitations, the maximum number of mobile devices that one Pi could support is 90:

$$Min (450, 100, 90) = 90$$

Experiment 3: Timeliness messaging

In real emergency situations, many community residents evacuate at the same time, using all the functions—including location tracking, connection, emergency messaging, and risk reporting—which will place great pressure on the Raspberry Pi network and affect the actual communication time. We conducted a simulated experiment of a real tsunami evacuation on the University of Pittsburgh campus to test the total time required for communication in the Pi network.

TABLE 7-5. RTT and packet success rate in Phone-Pi case

Load (%)	RTT (SD)	Success Rate (%)
10	47.972 (279.820)	99.705
30	43.578 (192.153)	99.445
50	58.824 (232.290)	99.215
70	70.968 (253.776)	99.228
90	122.073 (960.857)	98.682

Scenario. The experiment set up ten Raspberry Pis to form a local communication network for a simulated community at the University of Pittsburgh, with two students, one as leader and one as follower in the network. The first Raspberry Pi is the Main Pi of the community and the Home Pi of both leader and follower, who connect to the local EOC through the internet. A distance of 30 meters is set between the neighbor Pis. In the experiment, we made the following assumptions:

1. Ten phones simulate one hundred people using the same method described in mobility connections management; we set the random area beside the Pis less than 30 meters to simulate population density.
2. Phones send tracking information to the current Pi every 10 seconds to generate environmental noise.
3. No more than three phones are allocated to the same Pi, each phone runs ten tracking threads, and communication traffic load is less than 30 percent to keep the Pi working smoothly.

When the tsunami warning is sent from the local EOC through the internet, the follower moves only when the phone receives the warning. The follower keeps walking in the Pi network to follow the default route, sending tracking information and receiving the response. The follower sends an emergency assistance request every 30 seconds using the default option to send a request. The leader responds immediately when the request is received; the leader reports the road segment at risk and gets detours during the evacuation. The smartphone logs the time stamp of all actions: warnings are sent and received; moves are started and stopped; and tracking reports emergency assistance requests, risk reports, and reports of receipt of detour information and arrival at shelter.

Results. The result shows that the default route from the follower location to the shelter is 150 meters, the follower uses 141 seconds, and average speed is 1.1 meters/second, which is used as the basic parameter (online appendix, figure 13a).

Mobility tracking

With environmental noise, the packet success rate is 60.48 percent, which means that to get one successful message, the user needs to send at least two messages: 700 milliseconds maximum RTT[131.704 ms(RTT) +

568.122 ms(SD)] gives the user a chance to send twice in 1 second. Without environmental noise, the packet success rate is 79.85 percent, which is 72.545 milliseconds (249.176) as RTT(SD). Thus, users could successfully update their location to their Home Pi in 1 second, which leaves more time for users to communicate with others when they move into the reliability Wi-Fi of the Pi.

Emergency assistance

The emergency assistance results show that the average processing time for the leader is 2595.344 millisecond. With and without environmental noise, the packet success rates are similar (57.262% and 61.29%, separately). The 30 percent traffic load will not greatly affect the Pi's function because the rate is affected by moving in a low Wi-Fi signal strength region. The packet could not send or receive messages, as discussed in figures 7-1 and 7-3. When moving away from the Home Pi along the Pi network, the distance (number of hops) will also affect the RTT. We considered two types of RTT: (1) communication RTT = (sender receives − sender sends) − (receiver send − receiver receives) to show the communication infrastructure capacity (technical performance) and (2) total RTT = follower sends request − receives response to show the technical + social performance, which includes the user's response time to operate the APP in the mobile phones. As shown in table 7-6 and the online appendix, figure 14, when the leader and follower are two hops away from each other, the communication and total RTT are 160.5 milliseconds (15.5) and 3,443 milliseconds (861). That is an average increase from 336 milliseconds and 311 milliseconds every two hops, to 2,896 milliseconds and 5,239 milliseconds in eighteen hops. When the leader and follower are 540 (18 × 30) meters away from each other, the one-way communication times are 1,448 milliseconds and 2,619ms, and the infrastructure is fast enough to support emergency communication and evacuation in the community.

Risk reporting and detour

The results show that the total evacuation time is 170 seconds and that the follower who receives the detour message and evacuates uses a total of 152 seconds. The total detour distance is approximately 155 meters, which means that the risk awareness and operation time is 170 − 152 = 18 seconds (figure 7-7). If the leader is in self-detour mode, the route

TABLE 7-6. RTT (ms) and packet success rate (%) with hops

Hops	Communications RTT (SD)		Total RTT (SD)	
2	160.5	(15.5)	3,443	(861)
4	227.6	(125.141)	2,130	(364.365)
6	597.261	(429.622)	3,308.87	(1,918.003)
8	1,112.125	(978.413)	3,576.25	(1,351.726)
10	1,545.68	(914.311)	3,676.88	(1,005.913)
12	1,752.231	(733.563)	3,821	(733.563)
14	2,046.25	(694.253)	4,440.167	(1,385.203)
16	2,135.6	(1,026.563)	5,134.2	(1,594.559)
18	2,896.167	(1,193.012)	5,239.167	(1,205.549)

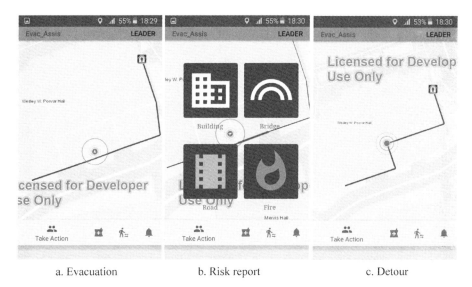

a. Evacuation   b. Risk report   c. Detour

FIGURE 7-7. Report and detour: (a) evacuation; (b) risk report; (c) detour.

bypasses the risk segment and then keeps following the old default evacuation route, and the total time is 213.321 seconds at a distance of approximately 200 meters, which loses 50 seconds compared with receiving the detour message from the local EOC.

Conclusion

This chapter proposes a sociotechnical infrastructure based on Raspberry Pi and mobile devices, which includes following methods:

1. A modified user datagram protocol (UDP) multicast mechanism to disseminate information, with each mobile device assigning a random transmission probability, to avoid information explosion.
2. A handover scheme with all Pi using the same SSID and all mobile devices using the same static IP to realize seamless handover when the mobile device is moving in the Pi network with low latency.
3. A Home Agent and care of address (CoA)–based mobile device tracking system. Each mobile device is assigned to a specific Home Agent Pi when registering in the Pi network, and the current Pi is assigned as the CoA to guarantee that information would be forwarded to the target mobile device.
4. An adaptive mobile device connection association mechanism to enhance stability and timeliness of the tracking and messaging process when the mobile device moves into weak Pi Wi-Fi signal strength regions.
5. A dynamic activity leader election mechanism to guarantee stable emergency social network connections between local government agencies and community residents at risk.

The basic functional experimental results show that using a measure of 30 meters as one hop, the Pi network could maintain a stable Pi-Pi TCP connection and Pi–mobile device UDP communication, with a Wi-Fi signal strength of approximately −70 and −80 dBm and RTT(SD) 11.177 milliseconds (20.318). The communication pressure experiment tested the limit of the number of mobile devices that one Pi could handle and reported that number to be approximately 90 devices. When the load increases to 90 percent, the sum of RTT(SD) between the Pi and the mobile device is 122.073 milliseconds (960.857), greater than the 1-second experimental limitation. When tested in a large-region Pi network with all handover, tracking, and connection functions working, the communication RTT is 160.5 milliseconds (15.5). When the leader and follower are two hops (60 m) away from each other, the time increases from 336 milliseconds for every two hops, 2,896 milliseconds (1193) to eighteen hops. That is, when the leader and follower are 540 meters away from each other, the one-way communication time is 1,448 milliseconds and the infrastructure is fast enough to support emergency communication. When users evacuate with an average speed of 1.1 meters per second, the tracking packet success rates between the mobile device and the Pi are 60.48 percent and 79.85 percent, 131.704 milliseconds (568.122) and 72.545 milliseconds (249.176) RTT(SD), with and without noise, respectively.

Emergency assistance average processing time for the leader is 2,595.344 milliseconds. With and without noise, emergency assistance packet success rates are similar (57.262% and 61.29%, separately).

The bowtie architecture collects, extracts, and integrates valid information from multiple sources into a common format and analyzes the information from the perspective of the whole community as a dynamic, adaptive system. It guides organizational action among multiple organizations, improving their capacity for communication and coordination while facing shared risk.

The findings from the study experiments demonstrate that this infrastructure could serve as a critical complementary part of the current Indonesia Tsunami Early Warning System to support community resilience. It would support the neighborhood preparedness groups (KSB; see chapter 2) as residents increase their situational awareness of risk by getting timely information from a stable communications infrastructure. Increased risk awareness among neighborhood groups builds collective risk awareness for the whole community through real-time communication, thereby enhancing collective action.

Use of the network of Raspberry Pi and mobile devices integrated in a bowtie architecture could effectively improve collective action in the early moments of a disaster. Policy makers at multiple levels could apply this evidence-based architecture to their decision support systems, combined with other information technologies, to improve the capacity of community residents and groups to manage shared risk more effectively. The results demonstrated in this field study will improve communication and coordination under uncertain conditions in context of disaster response.

References

Ai, F., L. K. Comfort, Y. Dong, and T. Znati. 2016. "A dynamic decision support system based on geographical information and mobile social networks: A model for tsunami risk mitigation in Padang, Indonesia." *Safety Science* 90: 62–74.

AlShamaa, D., F. Mourad-Chehade, and P. Honeine. 2018. "Mobility-based tracking using WiFi RSS in indoor wireless sensor networks." In *2018 9th IFIP International Conference on New Technologies, Mobility and Security (NTMS)*, 1–5. IEEE.

Arpaci-Dusseau, R. H., and A. C. Arpaci-Dusseau. 2018. *Operating Systems: Three Easy Pieces*. Arpaci-Dusseau Books LLC.

Biswas, S. and R. Morris. 2004. "Opportunistic routing in multi-hop wireless networks." *ACM SIGCOMM Computer Communication Review* 34, no. 1: 69–74.

BMKG. 2012. *Tsunami Early Warning Service Guidebook for InaTEWS*, 2nd ed. Jakarta: Government of Indonesia. Supported by GIZ-IS Protects (Project for Training, Education, and Consulting for Tsunami Early Warning Systems).

Boukerche, A., and A. Z. Zhang. 2008. "A hybrid-routing based intra-domain mobility management scheme for wireless mesh networks." In *Proceedings of the 11th International Symposium on Modeling, Analysis and Simulation of Wireless and Mobile Systems*, 268–75. Association for Computing Machinery.

Boukerche, A., Z. Zhang, and S. Samarah. 2009, July. "A WiFi-based wireless mesh network with inter-domain mobility management." In *2009 IEEE Symposium on Computers and Communications*, 857–62. IEEE.

Comfort, L. K. 2005. "Risk, security, and disaster management." *Annual Review of Political Science* 8, 335–56.

Comfort, L. K. 2007. "Crisis management in hindsight: Cognition, communication, coordination, and control." *Public Administration Review* 67: 189–97.

Csete, M., and J. Doyle. 2004. "Bow ties, metabolism and disease." *TRENDS in Biotechnology* 22, no. 9: 446–50.

Ferdoush, S., and X. Li. 2014. "Wireless sensor network system design using Raspberry Pi and Arduino for environmental monitoring applications." *Procedia Computer Science* 34, 103–10.

Glass, R. J., W. E. Beyeler, A. L. Ames, T. J. Brown, S. L. Maffitt, N. Brodsky, P. D. Finley, et al. 2012. "Complex Adaptive Systems of Systems (CASoS) Engineering and Foundations for Global Design." Albuquerque, NM: Sandia National Laboratories, SAND Report, SAND2012-0675. January.

Han, J., E. Haihong, G. Le, and J. Du. 2011. "Survey on NoSQL database." *In 2011 6th International Conference on Pervasive Computing and Applications*, 363–66. IEEE.

Hossmann, T., F. Legendre, P. Carta, P. Gunningberg, and C. Rohner, 2011. "Twitter in disaster mode: Opportunistic communication and distribution of sensor data in emergencies." In *Proceedings of the 3rd Extreme Conference on Communication: The Amazon Expedition*, 1. Association for Computing Machinery. ExtremeCom2011, Manaus, Brazil, September 26-30.

Hu, Y., and R. Huang. 2015. "Development of weather monitoring system based on Raspberry Pi for technology rich classroom." In *Emerging Issues in Smart Learning*, 123–29. Berlin: Springer.

Iyer, A., and J. Iyer. 2009. "Handling Mobility Across Wi-Fi and WiMAX." *Proceedings of the 2009 International Conference on Wireless Communications and Mobile Computing, Leipzig, Germany, June 21–24*, 537–41. New York: Association for Computing Machinery.

Kahneman, D. 2011. *Thinking, Fast and Slow.* New York: Farrar, Straus and Giroux.

Khelil, A., F. K. Shaikh, A. A. Sheikh, E. Felemban, and H. Bojan. 2014. "Digiaid: A wearable health platform for automated self-tagging in emergency cases." In *2014 4th International Conference on Wireless Mobile Communication and Healthcare-Transforming Healthcare through Innovations in Mobile and Wireless Technologies (MOBIHEALTH)*, 296–99. IEEE.

Martínez, F., Montiel, H., and H. Valderrama. 2016. "Using embedded robotic platform and problem-based learning for engineering education." In V. L. Uskov, R. J. Howlett, and L. C. Jain. *Smart Education and e-Learning 2016*, 435–45. Cham, Switzerland: Springer International Publishing AG.

Mileti, D. S., L. Peek, and P. Stern. 2002. "Understanding individual and social characteristics in the promotion of household disaster preparedness." In *New Tools for Environmental Protection: Education, Information, and Voluntary Measures*, 127–32. Washington, DC: National Academies Press.

Nakauchi, K., and Y. Shoji. 2016. "vBS on the move: Migrating a virtual network for nomadic mobility in WiFi networks." In *2016 IEEE Conference on Computer Communications Workshops (INFOCOM WKSHPS)*, 277–82. IEEE.

Paramanathan, A., P. Pahlevani, S. Thorsteinsson, M. Hundeboll, D. E. Lucani, and F. H. Fitzek. 2014. "Sharing the pi: Testbed description and performance evaluation of network coding on the Raspberry Pi." In *2014 IEEE 79th Vehicular Technology Conference (VTC Spring)*, 1–5. IEEE.

Postel, J. 1980. "User Datagram Protocol." *Internet Engineering Task Force* (August). https://www.rfc-editor.org/rfc/rfc768.

Rittel, H. W., and M. M. Webber. 1973. "Dilemmas in a general theory of planning." *Policy Sciences* 4, no. 2, 155–69.

Simon, H. A. 1996. *The Sciences of the Artificial*, 3rd ed. MIT Press.

Tsunami Research Center. 2010. "The 25 October 2010 Mentawai earthquake and tsunami." University of Southern California Tsunami Research Center. http://www.tsunamiresearchcenter .com/news/the-25-october-2010-mentawai-earthquake-and-tsunami/.

Yang, W. H., Y. C. Wang, Y. C. Tseng, and B. S. P. Lin. 2009. "An energy-efficient handover scheme with geographic mobility awareness in WiMAX-WiFi integrated networks." In *2009 IEEE Wireless Communications and Networking Conference*, 1–6. IEEE.

Zhou, M., Z. Tian, K. Xu, X. Yu, X. Hong, and H. Wu. 2014. "SCaNME: Location tracking system in large-scale campus Wi-Fi environment using unlabeled mobility map." *Expert Systems with Applications*, 41, no. 7: 3429–43.

Zhou, M., K. Xu, Z. Tian, H. Wu, and R. Shi. 2015. "Crowd-sourced mobility mapping for location tracking using unlabeled Wi-Fi simultaneous localization and mapping." *Mobile Information Systems* 2015, article ID 416197. https://doi.org/10.1155/2015/416197.

Chapter 8

Real-Time Seafloor Tsunami Detection and Acoustic Communications

Lee Freitag, Keenan Ball, Peter Koski, James Partan, Sandipa Singh, Dennis Giaya, and Kayleah Griffen

An early warning system for near-field tsunamis consisting of a seafloor detector and an undersea wireless communications capability has been developed for initial testing and evaluation in the Mentawai Basin south of Sumatra, Indonesia. The sensor module on the seafloor consists of a precision pressure gauge, an event detector, and an acoustic modem. A complementary acoustic modem that acts as the receiver is placed at the end of a subsea cable, which provides power and hard-wired communications to shore to complete the link. The base station on shore includes a backup power system and satellite data link to the Indonesian Tsunami Early Warning System (InaTEWS), where the data are incorporated into existing warning capabilities. This new approach, still in development and undergoing initial trials, is a prototype for hybrid tsunami warning systems that provide additional options to the detection of deep-ocean assessment and reporting of tsunami (DART) buoys or cabled systems. The advantage of the acoustically linked system over the DART is that it does not require the surface buoy that is prone to damage from storms or vandalism. An advantage over purely cabled systems is that it reduces the amount of cable to 7 kilometers in the test case described here, rather than 30 kilometers as would otherwise be required, resulting in reduced capital costs. This approach is not appropriate for all areas threatened by near-field tsunamis, however, as it takes advantage of acoustic propagation enabled by warm tropical surface water and proximity of deep water to shore.

The Sumatra prototype was initially tested in 2016 to establish the feasibility of the long-range horizontal acoustic link. A complete version, configured to detect pressure anomalies and transmit short messages over an acoustic path of 25 kilometers, was then deployed in early

2020. In addition, the system transmits the local pressure record with the tidal signal each day to confirm that it is operational. The project serves to evaluate feasibility and applicability to other areas of Indonesia, particularly on the Indian Ocean, which is threatened by tsunamis generated by earthquakes at the Sunda megathrust. This chapter describes the motivation for the approach, results of the initial acoustic testing, and implementation of the complete prototype in 2020.

Mitigating Tsunami Hazards

Tsunamis from subsea earthquakes and landslides threaten many coastlines of the world. The nature of those threats varies widely depending on location, however, and thus the approach to mitigating the hazards from tsunamis also varies. In the *far field*, hundreds to thousands of kilometers from the epicenter of a subsea earthquake, there may be many hours between the initial event and the arrival of a tsunami wave on shore. In the *near field,* however, there may be only 15 to 30 minutes from when the earthquake occurs to when the wave arrives. It should be noted that estimates of tsunami magnitude and arrival time on shore are made using shore-based and subsea seismometer measurements, although the estimates made in the first few minutes of an earthquake may be approximate. Further, some shallow subduction zone earthquakes may be quite moderate but highly tsunamigenic, with only modest locally felt shaking that may not be indicative of the potential wave height and subsequent inundation (Polet and Kanamori 2000). As a result, the most accurate way to detect and predict a tsunami is via direct measurement of the wave as it passes overhead a seafloor pressure sensor.

The need for direct measurement motivated the development of the US DART buoys (Bernard and Meinig 2011), which consist of two parts. One part is a seafloor unit with a very accurate quartz crystal sensor that can measure pressure to a fraction of a centimeter; the other is a buoy with a modem receiver and a satellite system for transmitting the warning back to shore. The seafloor sensor and the buoy are connected by a short-range (5 to 7 km) acoustic communications system that allows the seafloor pressure signal to be monitored in near real time. Under normal circumstances, the sensor reports at a very low rate, but upon detecting a pressure anomaly, it changes to a rapid reporting mode and sends back data continuously for several hours. The DART buoys are located around the United States and provide warnings for incoming tsunami waves for the West Coast and Alaska (Gonzàlez et al. 2005), as well as in the Atlantic.

The approach presented here was inspired by the DART buoy system but tailored to the unique characteristics and constraints of the tsunami warning situation in Indonesia, where use of buoys has proven to be problematic (Data Buoy Cooperation Panel and International Tsunameter Partnership 2011). As a result of the acknowledged problem of maintaining buoys, particularly those close to shore and prone to interference, multiple approaches were considered, including undersea cable systems and use of subsea acoustic networks.

In Indonesia, the Agency for the Assessment and Application of Technology (BPPT) experimented with short, cabled pressure sensors, using a relatively low-cost cable deployed from an oceanographic research vessel to avoid the use of expensive telecommunications cable-laying ships. This approach can be quite inexpensive and allows for cables from 5 to 10 kilometers to be deployed near shore. In addition, in the United States, researchers at the University of Pittsburgh, including an Indonesian engineer from BPPT enrolled as a graduate student, begin considering use of acoustic modem sensor networks. The initial research into network design envisioned multiple sensors and relay nodes to provide redundancy in both detection and data transmission (Comfort et al. 2012; Xerandy et al. 2015). Further exploration into options for implementation of the sensor network brought in researchers from the Woods Hole Oceanographic Institution (WHOI) in Massachusetts, who had done work on longer-range acoustic links, which could potentially simplify the system (Freitag 2000). The synergy of these efforts has led to the work described in this chapter, which focuses on the acoustic link and the subsea sensor, and the other chapters of this book.

The specific phenomenon that the acoustic link demonstrated for this project exploits is refraction (bending) of sound waves. In water less than 2,000 meters deep, acoustic rays will refract from the warmer, faster water near the surface and return to the bottom at ranges of 20 to 40 kilometers, depending on depth and gradient of the surface layer. In regions where the surface waters can be cold during the winter, refraction may not occur, and the surface-refracting phenomena are seasonal. In contrast, in tropical areas of the world, the warm surface layers are well established and persist throughout the year, guaranteeing that rays will refract and not reflect.

Due to its location, refracted bottom-to-bottom communications is practical in areas such as the Mentawai Basin between West Sumatra and the Mentawai Islands of Indonesia. The depth of this area is approximately 1,700 meters at its deepest, and the bathymetry rises steeply to the west toward Siberut Island and more gradually to the east toward the city of Padang (see the map in the preface), which is threatened by tsunamis created by earthquakes caused by pent-up energy in the Sunda megathrust. Thus, the team envisioned and then explored use of long-range bottom-to-bottom acoustic communications to exploit the use of low-cost short cables and implement the hybrid near-field tsunami warning system as shown in figure 8-1.

This chapter is organized into three specific sections. The first provides the background for the acoustic-based system and describes a feasibility test that was performed in 2016 to confirm the assumption about acoustic propagation and communications; the second describes the processing method used for the pressure sensor data and how detection events are identified and then telemetered over the acoustic link; and the third describes the initial deployment of the complete system in early 2020, which, even though it was not brought to an operational state, demonstrated the basic principles of functionality

Background and Feasibility Test

Underwater networks for remote sensing applications have been a topic of research for many years in the underwater acoustic communications

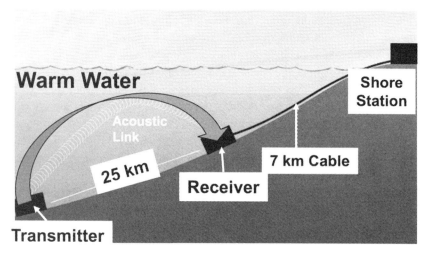

FIGURE 8-1. Hybrid near-field tsunami warning system.

community. The challenges, solutions, and myriad research areas have been well summarized in a number of articles on sensor networks (Heidemann et al. 2012; Proakis et al. 2001). This work builds on earlier work supporting seafloor networks, such as those described in several articles on the US Navy's Seaweb concept (e.g., Rice and Green 2008), and observatory systems developed and demonstrated by the authors in 2010 at the Monterey Bay Aquarium Research Institute undersea observatory (Freitag et al. 2010). The previous work utilized systems operating at 10 kilohertz or higher, which limited the possible range of the acoustic modems to less than 5 to 10 kilometers, depending on conditions and source-receiver geometry.

For the near-field tsunami warning application, the potential distance between sensors is relatively large, on the order of tens of kilometers, and the distance to shore even longer, thus motivating a modem physical layer that is lower frequency—although not so low as to excessively constrain the communications bandwidth. For this feasibility study, a carrier frequency of approximately 3 kilohertz was chosen based on previous experiments done at 2.25 kilohertz by the authors in 1996 that demonstrated ranges up to 44 kilometers (Freitag et al. 2000). The frequency was selected as a compromise between cost (and transducer size) and range, with transducers readily available from several vendors in the 3 to 4 kilohertz range, such as Geospectrum in Canada and ITC in the United States.

The goal of the experiment was to explore the point-point performance with a design based on the WHOI Micromodem, which is a family of underwater acoustic modems, so that a system could be put into service as soon as possible. Although networking is not explicitly addressed in the work presented here, the rationale for the test was to provide input for a network design that can be accomplished at minimum cost for InaTEWS.

The ultimate use of the acoustic communications network is for detection of near-field tsunami waves and transmission of a detection alert and a portion of the pressure data to shore, both in near real time. One challenge in implementing such a network is unambiguously identifying the pressure wave from coincident strong ground motion on a precision pressure sensor, such as the Paroscientific (US) Digiquartz unit. The real-time identification of a near-field tsunami was beyond the scope of this first phase of the project, but work on this area is of interest because a number of areas in the world are subjected to the threat of a near-field event where the warning time is very small (Schindelé et al. 1995).

Ultimately, the system employed a precision sensor such as described in Paros and colleagues (2012), with onboard processing to separate the tsunami signal from ground motion and a detection method similar to that developed by the National Ocean and Atmospheric Administration (NOAA) in the United States for the DART buoy (Bernard and Meinig 2011). A broadband Paroscientific pressure sensor operating at 20 hertz was deployed as part of this experiment to gather background data for development of the detection algorithm, with the data being saved for later analysis.

Experiment overview

The acoustic testing was accomplished using a local vessel, the *KN MUCI*, made available by the Indonesian government department responsible for buoy maintenance in West Sumatra. The test was conducted in March 2016 from a port south of Padang where mobilization and demobilization took place, and it was done in cooperation with the University of Andalas in Padang (Professor Febrin Ismail) and the BPPT (Iyan Turyana and X. Xerandy). The test area, shown in the map in the preface, was selected because it offers close access to the deep-water area of the Mentawai Basin and because it has been identified as a potential shore site for the acoustic receiver, cable, and shore station. The acoustic systems were deployed very close to the bottom (4 m) at depths that varied from 900 meters (closest to shore) to 1,700 meters (the typical depth of the basin).

Source-receiver ranges varied from 5 to 33 kilometers, with one pair of modems used for bidirectional testing and a third in receive-only mode. Three recorders were used as well, with one colocated with the receive-only modem and the other two at independent locations to collect additional data. A subset of the data that was collected is reviewed in the results section of this chapter.

Sound speed profile and acoustic propagation

The sound speed profile corresponding to the temperature and salinity measured by free-falling a conductivity, temperature, and depth set of instruments colocated with one of the acoustic systems deployed to 1,700 meters is shown in figure 8-2a. The warm layer at the surface is clearly evident in the fast sound speed between 0 and 200 meters deep. The sound-speed minima is at approximately 1,000 meters, the same depth as in much of the Pacific, although it is not as slow because the water is significantly warmer in this region (6°C as opposed to 4°C in other tropical parts of the Pacific).

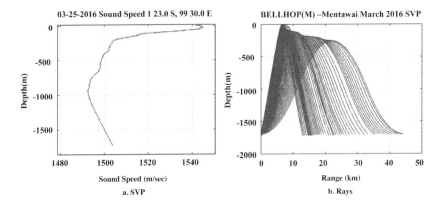

FIGURE 8-2. (a) CTD profile showing the sound speed with respect to depth taken during the experiment. The (x) marks show the gridded profile used in propagation modeling. (b) Rays corresponding to the measured surface velocity program. For simplicity of viewing, only a subset of rays is shown.

Using ray-tracing software, the propagation paths for the area were calculated and mapped as shown in figure 8-2b. The plot shows how the acoustic rays travel through the ocean with the measured sound-speed profile. The rays that start out at high angles toward the surface intersect with the surface and are then reflected back to the bottom. This behavior continues until approximately 15 kilometers from the source. After that distance, the rays are all refracted and do not intersect with the surface. Direct path rays from bottom to bottom may exist at close range (several kilometers) even though the sound speed profile refracts upward. The upward refraction, which causes the rays to bend away from the bottom, is what typically limits bottom-to-bottom communications in deep water.

At ranges from approximately 15 to 45 kilometers, the rays that leave the source at angles such that they do not interact with the bottom will be received at bottom receivers without interaction with the surface. Thus, they undergo no boundary loss, only range-dependent spreading and frequency-dependent absorption. Using a 3.5 kilohertz carrier frequency, the estimated range for the acoustic system was 20 to 30 kilometers at a source level of 185 decibels re micropascal (the standard unit used for underwater sound level).

Communications and recording setup

The WHOI Micromodem is an evolving research, development, and production acoustic communication system that was developed in the late 1990s and most recently updated in 2010 (Gallimore et al. 2010). Since then, it has had many functional updates for precision timing, sensor interfaces, and new transducers. Over the years, it has been used in many demonstrations, and it forms the core of the acoustic communications capability in the Remote Environmental Monitoring Units, or REMUS, class of unmanned underwater vehicles. Recent work has included adapting it to a wider range of transducers and thus carrier frequencies, including the 3 kilohertz systems described here and at lower frequencies, such as 900 hertz for a wide-area Arctic acoustic communications and navigation system (Freitag et al. 2015).

The onboard processing utilizes a multichannel decision-feedback equalizer based originally on the work of Stojanovic and colleagues (1994) but with a number of modifications for real-time implementation, such as an adaptive least-mean-square update algorithm. An example of a previous performance measurement experiment involving this algorithm is described in Freitag and Singh (2009).

Acoustic signals

To measure the performance of the proposed system in terms of reliability and data rate, a number of signals were transmitted. Two signal bandwidths, and thus symbol rates, were used, 625 and 1,250. The only significant difference between the two is that the lower bandwidth has approximately double the energy per symbol. Three different modem rate settings were also used. The different rates employ different levels of error-correction coding, with rate 1 being the lowest and rate 5 being the highest.

Data rates from approximately 60 to 700 bits per second were tested during the experiment, providing a factor of ten in variation of transmitted data rate. One hundred bytes of user data were transmitted in each packet. The latency of the receiver is low, approximately 1 second. Thus, the time from the start of the packet to the time when the data are received and available to the user, or ready for retransmission to the next node, is simply the duration of the packet plus 1 second. This information is important for real-time use in the tsunami warning system application.

The packets consist of a frequency-modulated sweep, followed by a short signal burst that is used to set the gain of the receiver, followed by 510 training symbols and then the acoustic data. Each packet contains a header that informs the receiver of the encoding of the data packet, which can be configured with different levels of error-correction coding depending on the reliability of the acoustic channel, specifically the signal-to-noise ratio (SNR). The results discussed below are tabulated in terms of the SNR.

Results

In this discussion, the results at ranges at 20 and 27 kilometers are deemed of most interest and are described in detail. Communications in the presence of a significant surface bounce were also feasible at close range (5 km), however, as shown in figure 8-3a.

Close-range impulse response (5 km). The channel response at close range demonstrates that a broad transmit beam broadcasts rays that travel directly and via surface bounce (figure 8-3a) to other seafloor receivers. The very long surface-bounce delay results in poorer performance than at ranges where only the near-surface *refracted* path is present. Thus, although it is interesting that the system can work with the reflections, this range is not relevant to the proposed long-range link.

Refracted path impulse response (25 km). At longer ranges, the impulse response becomes quite simple: it is a refracted path with little discernable

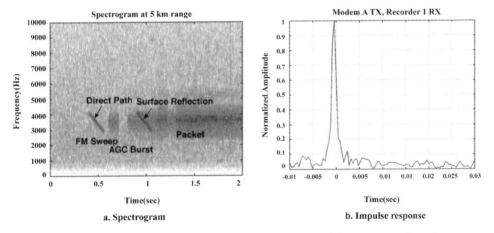

FIGURE 8-3. (a) Spectrogram showing surface bounce at 5 kilometers. (b) Impulse response for a single channel at 25 kilometer range. The single peak shows that there is just one path from source to receiver, validating the assumptions about simple acoustic propagation in this environment.

spreading—at least at the bandwidth used for this experiment, which was limited to 1,250 hertz. Figure 8-3b shows an example, obtained from one of the recorders over a 25 kilometer path. All four hydrophone channels look very similar, with channel-to-channel delay commensurate with the angle of arrival and array element spacing.

Communications results

System performance is dictated by a combination of input SNR, which is measured in decibels, plus the local multipath due to reflections or refraction and the rate of change of that multipath. For a fixed geometry with stationary source and receiver, the multipath changes are a function of variability within the water column, which away from the surface layer is typically minimal. We observed variations in the input SNR, however, indicating that at the test ranges the acoustic channel does vary over short time periods. Figure 8-4a shows an example of the input SNR over the two experimental periods that correspond to the 20 and 27 kilometer range.

The output SNR is a function of the input SNR, which bounds the overall performance, but it is also affected by the amount of multipath

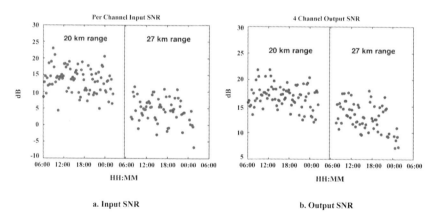

a. Input SNR   b. Output SNR

FIGURE 8-4. (a) Input signal-to-noise ratio (SNR) at modem A with modem B transmitting at two different ranges over the two days of testing at 625 symbols per second. The average input SNR is 13.5 decibels (dB) at 20 kilometers and 4 dB at 27 kilometers. (b) Output SNR after four-channel equalization. The mean output SNR is 16.8 and 12.9 dB, respectively, for the two different ranges. Results for the highest data rate (rate 5) are shown, which has a burst data rate of 350 bits per second. One packet has a bad CRC (dark gray).

and its rate of change, which the adaptive equalizer must estimate and track. In addition, the output SNR after four-channel equalization is typically higher than the input SNR due to the processing gain. Figure 8-4b shows the output SNR that results when all four hydrophone channels are processed by the adaptive equalizer. The burst data rate for this example is 350 bits per second, with the error-correction coding successful in all but one case where the output SNR is getting close to the limit of approximately 7 decibels, where occasional bad frames begin to occur. The success rate is better than 98 percent in this case.

Conclusions regarding the experiment

The experiment demonstrated that direct path communication is feasible from sensors located on the seafloor to cabled receivers or relay nodes located at moderate distances, approximately 30 kilometers. These results are predicated on two conditions: (1) that the surface water is warm and thus refracts the sound away from the surface before it can scatter and (2) that the bottom depth (and thus the source depth) is shallower than in typical full-ocean conditions where convergence zones are formed at the 50 to 60 kilometer range.

The demonstrated data rate of 350 bits per second with very high reliability shows that a tsunami alert system could be implemented with one to several hops and typically will have a very low probability of requiring a retransmission to correct a bad frame. Higher data rates are feasible as well, with somewhat lower overall reliability. For an acknowledgment channel for the return link from shore to the remote sensor, the lowest data rate of 60 bits per second can be used to ensure that control messages are received with high reliability.

Tsunami Signal Detection Using a Pressure Sensor

Declaring a tsunami detection is a time-critical event with real-world implications for false positives and especially grave consequences for false negatives. As such, the ultimate decision is best left up to an expert using sophisticated modeling software capable of fusing measurements from a diverse array of sensors. There are specific challenges when integrating a remote underwater pressure sensor connected by an acoustic link into such a system. Because the quality and relevance of measurements provided by such a sensor are potentially very high, the challenges of operating a sensor on a finite and limited power budget with a high latency and low bandwidth acoustic link must be considered.

The standard method for tsunami detection is through use of a precision pressure sensor on the seafloor. The approach was developed and reduced to practice by researchers at NOAA, which resulted in the DART buoy design. Other similar buoys have been developed and deployed by other countries. A typical signal from a tsunami traveling in deep water may be only a few centimeters in height, whereas the tidal signal is often 1 or more meters. Thus, the standard method for detection involves estimating the tidal signal and removing it, then triggering at a threshold of approximately 1 centimeter. When triggered, the DART systems start the high-frequency sampling and reporting mode to capture and transfer the data back to shore. After several hours, they return to their normal operating mode. In addition to the data transmitted when the threshold is exceeded, the seafloor node associated with a DART buoy also sends back data at regular intervals to show that the system is functional. This process results in transmission of the local tide.

In the case of the system to be employed in the Mentawai Basin, the pressure sensor is located close to the source of the earthquake. The pressure sensor will thus pick up the ground motion, as well as the signal generated by movement of the ocean surface in response to motion of the seafloor. The ground motion signal has different frequency content than that of the tsunami wave, but there is overlap between them, thus making the detection problem more difficult than in deep water.

In the Mentawai Basin, the detection occurs in the *near field*, whereas when the sensor is far from the tsunamigenic earthquake, it is in the *far field*. The detection of a tsunami in the near field is a problem that is present in multiple locations, including off Japan, Chile, and the West Coast of North America. Data recordings with coincident bottom movement due to the seismic activity and a tsunami are rare, but there have been fortunate situations when sensors have recorded both the seismic signal and the tsunami wave some hours apart, which allows for comparative analysis. To illustrate the detection of a tsunami signal, data from several locations were used to examine different options for filtering and setting a detection threshold, including from the 2012 Haida Gwaii (British Columbia, Canada) earthquake and tsunami (Sheehan et al. 2015). The earthquake occurred on October 28, 2012, off the west coast of Canada with a magnitude of 7.8, and the resulting seafloor uplift generated a tsunami measuring between 3 and 13 meters on shore (Leonard and Bednarski 2015).

The processing that was performed began with decimating the data originally sampled at 125 hertz to the much lower rate of 0.1 hertz, with one sample every 10 seconds. The next step was removing the tidal signal with a high-pass filter and then filtering again in the tsunami band. The frequency band for the tsunami signal was set to be from 2 to 90 minutes, and the data were filtered within that band. The resulting signal was then examined to determine how to set a detection threshold that will trigger the warning. As shown in figure 8-5, the filter removed much of the initial signal from seismic motion, but it is also clear that the tsunami signal, which comes much later, may be approximately the same amplitude as remnants of the motion signal that overlap into the tsunami band. In the case of a near-field event, the tsunami signal will be significantly higher, but the motion signal will be stronger as well. This approach demonstrates the issues associated with sensing for tsunami signals received on pressure sensors in the near field, particularly when processing and a detection decision must be made using a low-power processor running in real time.

The wireless underwater node operates on battery power and thus, to maximize the operational time of the node, each subsystem must be optimized for power consumption. The precision pressure sensor (manufactured by Paroscientific), although capable of sampling at much higher rates, is sampled once per second to save energy. The 1 hertz data cannot be streamed to shore for analysis because of the high power cost associated with acoustic transmissions, as well as the limited bandwidth of the link. Instead, the local detector monitors the data stream locally and divides the data into one of three categories: tidal signal, interesting seismic events, and potential tsunami events. Each class of data is determined by the onboard microprocessor that then uses it to generate one of three possible reports: the daily report, the seismic event report, and the tsunami event report. For the purposes of the implementation for the prototype system, the processing was simplified to use a running average. Although this process may not provide the best possible prefilter for tsunami detection, for the first deployment of the detection system it was decided to keep the processing as simple as possible. The details of the processing and the resulting reports that are generated for transmission by the modem are described below.

All sea-surface-height measurements not belonging to the seismic or tsunami category contribute to the tidal signal in the daily report. This report includes the time and a thirty-six-point representation of the sea surface height over the last 24 hours, with each point containing the mean

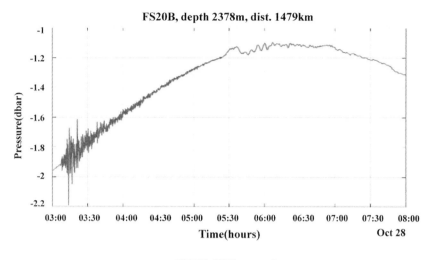

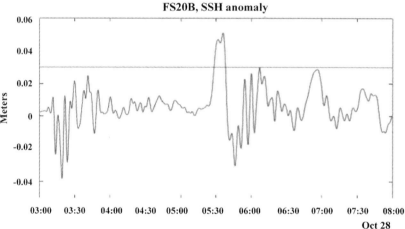

FIGURE 8-5. (top) The pressure signal from sensor FS20B, located 1,479 km from the epicenter at a depth of 2,378 meters. (bottom) The pressure signal after removal of the mean and filtering. The ground motion signal is still strong, apparent in the data record, whereas the tsunami occurs two hours later.

and standard deviation of the signal in a 40-minute window. Ultimately, this information allows for an accurate re-creation of the tidal signal at the exact location of the sensor and, over the course of multiple deployments, provides a dataset that can be used to predict the tide forward accurately. The daily reports are only 93 bytes long.

Seismic events and tsunami events are detected in real time by the microprocessor, and reports are sent to shore when events are detected. The detection algorithm is identical in both cases but uses different thresholds. Sea-surface-height measurements from the precision pressure sensor are first decimated from 1 to 0.1 hertz. After decimation, the mean of the most recent 10 minutes of samples (sixty samples) is computed. The windowed mean serves to filter out the high-frequency seismic signal that typically dwarfs the much lower frequency tsunami signal in magnitude. The value at the middle of this window is compared to the mean of the window, and the difference between the two events is the threshold used for comparison. A deviation of more than 1 centimeter suggests an interesting seismic event and begins to populate a seismic event report. A deviation of more than 3 centimeters suggests a potential tsunami event, and the system begins buffering the data to report the tsunami detection.

The seismic event report is sent after an additional twenty-four samples are collected, which is 4 minutes after the detection. It contains the time stamp and a compressed representation of the sixty most recent sea-surface-height measurements sampled at 0.1 hertz and is effectively a snapshot of the signal window that includes the most recent 6 minutes of the original window processed at the time of detection as well as the 4 new minutes of data up until the present. During this waiting period, the detector is still running, and if the tsunami threshold is exceeded, the device switches immediately into reporting a tsunami detection. The seismic event report is 73 bytes long.

Similar to the seismic event report, the tsunami event report contains the time stamp and a compressed representation of the sixty most recent sea-surface-height measurements sampled at 0.1 hertz, and it is a direct snapshot of the processing window at the time of detection. In addition, it contains twelve points summarizing the last 8 hours of tidal history at the sensor. This report is sent to shore immediately using the acoustic modem. Once the system has detected a potential tsunami, it continues reporting new data each minute for 12 minutes. The layout of each report is identical, and the reports' contents are overlapping, ensuring that the most recent available data are sent to shore to inform the decision-making process. The system also provides redundancy over the acoustic link if portions of an acoustic packet could not be decoded at the receiver. The tsunami event report is 97 bytes long.

The detection algorithm on the microprocessor, although simplistic, allows for relevant data to be transmitted to shore in a timely manner along with the necessary contextual information for a more sophisticated system to make a final determination. The accepted standard for detection is a 3 centimeter deviation in sea surface height. Although the onboard microprocessor can approximate this determination, the system is designed so that all the necessary information for a more precise estimate is provided to processing centers on shore.

The historical tide data provided daily generates an accurate tidal model at the location of the sensor. The most recent 8 hours of tidal data are sent with the tsunami report to fill in any gaps between the last daily update and the time of the tsunami detection. The time period of 8 hours sampled with 40-minute resolution provides data to account for deviations in the sea surface height due to atmospheric variations. Combined with the historical tidal signal, it should be possible to predict the nominal sea surface height at the sensor. Finally, each report contains the most recent 10 minutes of the sea surface height at 0.1 hertz. When compared to the nominal sea surface height, the tsunami signal should be apparent on this timescale. When considered in conjunction with other wireless tsunami sensor nodes, tide gauges, and a system of seismic sensors, a more accurate tsunami prediction can be made.

Initial System Deployment and Demonstration

The prototype near-field tsunami warning system designed to be deployed in the Mentawai Sea near Siberut Island consists of two components: (1) a remote unit on the ocean bottom to sense the tsunami signal and transmit the data using an acoustic modem and (2) a receiver that is cabled to shore on Siberut. The ocean bottom unit includes the high-resolution pressure sensor and processor described above. When a warning threshold is exceeded, the remote sensor subsystem will use the acoustic modem to transmit compressed data over the acoustic modem over a distance of approximately 25 kilometers. The receiver, at the end of a cable to shore, will forward that data over a satellite link for further processing and distribution within InaTEWS.

The work planned for late in 2019 included the integration of the underwater systems developed by WHOI with the cabled hardware and shoreside telemetry system developed by the Indonesian Agency for the Assessment and Application of Technology (BPPT). The work to be performed by BPPT was partially supported by the grant from the Swiss Re Foundation, administered through the University of Pittsburgh. After the integration was completed, the whole system—both the remote sensor and the cabled modem and the cable and shore station—was deployed near Siberut Island. To capture the nature of the work and the challenges that were involved in the effort, the steps and major events that were involved in the integration and deployment are described in this section.

Equipment

The equipment falls into two categories: one for connection to the cabled system and the other for the remote ocean bottom unit. All the equipment is designed and fabricated by WHOI. Components that are commercially sourced include the pressure sensor (from Paroscientific), the floats, and a few other items. Most of the other components that make up the completed system were designed by WHOI and fabricated either at WHOI or by suppliers.

Ocean bottom unit. The remote unit is completely self-contained and includes an instrument frame that rests directly on the seafloor with the pressure sensor, a cable that goes up to the acoustic modem, and flotation to suspend the modem in the water above the sensor, as shown in figure 8-6b. The instrument frame was initially designed in 2015 and then used in 2016 for the feasibility test. It was recovered from the seafloor by BPPT with support from Andalas University and sent to the BPPT facility, where it was refurbished in November 2019 by WHOI hardware engineers.

It is important that the sensor be on the seafloor in a separate structure so that it is not affected by mechanical noise due to cable strum created by currents. The acoustic modem, however, needs to be off the bottom so that there is a clear line of sight to the cabled system and it is not blocked by any nearby seafloor obstructions. The cable between the two is 15 meters long, and the breaking strength of the cable is 4,500 pounds.

There are two titanium pressure housings—one is located on the ocean bottom unit, and the other contains the acoustic modem—and each has its own batteries. The detection processor is located in the housing that contains the battery for the pressure sensor. Communication between the detection sensor and the acoustic modem is done with electrical conductors in the 15-meter-long cable. The acoustic modem is normally asleep and expending no energy. The detection processor wakes it up when an alert occurs or when regularly scheduled system status transmissions are to be performed.

Cable end unit. The unit at the end of the cable consists of two main components. One is supplied by BPPT as part of the cable power and

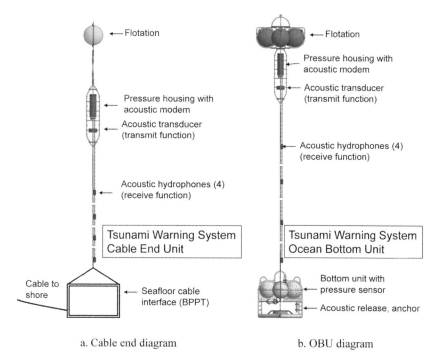

a. Cable end diagram   b. OBU diagram

FIGURE 8-6. (a) Cable end unit for installation at the end of the subsea cable to shore. (b) Tsunami system remote ocean bottom unit with pressure sensor and acoustic modem.

telemetry system, and the other is provided by WHOI and includes the acoustic modem receiver system.

The unit is very similar to the ocean bottom unit but without the frame that rests on the bottom with the pressure sensor, as shown in figure 8-6a. The acoustic modem will be provided power from the cable, and the communications will be done via the cable as well.

Although the function of the cable end unit is primarily to receive data from the ocean bottom unit, it can also send commands or acknowledgments. In fact, both modems are identical, and the link is symmetric. Although the remote unit will be powered off almost all the time, the modem at the cable end will be powered all the time and be able to receive signals from the remote unit whenever they are sent.

Field work completed

The initial WHOI team of mechanical engineers worked at BPPT from November 11 to November 15, 2019. During that time, the batteries were assembled in the housings and the electronics installed and tested. Work was also done to ensure that the power draw by the modem at the end of the cable was low enough to be supported by the shore system, which was done by reducing the maximum draw to approximately 6 watts. At the end of the integration period, the equipment was ready for final installation of software for the detector subsystem, assembly, and final test prior to deployment.

The second WHOI teamincluding hardware and software engineers arrived in Jakarta on December 16, 2019, and worked at the BPPT facility in Serpong. Work performed during that time included installing and testing a higher voltage converter for powering the acoustic system and installing the fiber to copper converters brought from the United States by WHOI. Work also continued on testing the tsunami signal detection software, integrating the detector with the acoustic modem, and finalizing the configuration of the detection subsystem and the data formats.

On December 21, 2019, Singh returned to the United States, and engineer Dennis Giaya joined the team at Serpong. Work continued there with final packing performed on Monday, December 23. On December 24, WHOI engineers traveled to Padang and then continued to the town of Tua Pajet on the island of Sipora on December 27, along with the majority of the BPPT engineers including Iyan Turyana and X. Xerandy, who led the at-sea and shore teams respectively. The BPPT team worked on preparing the Sipora cable on the December 28 and 29, which was then successfully laid on December 30. The Sipora cable is 7 kilometers long.

The changeover of the equipment for the Siberut cable was performed on December 31, 2019, and January 1, 2020, and the deployment vessel departed Tua Pajet at 1 a.m. on January 2 destined for the offshore site where the instrument frame with the ocean bottom unit with the pressure sensor was to be deployed. The instrument was deployed later that morning in a depth of approximately 1,700 meters at a location 30 kilometers offshore and 25 kilometers from the planned location for the cable end. The ship then docked on Siberut Island that evening. On January 3, the cable was laid with the modem and cable termination in 700-meter-deep water. The cable was paid out too quickly during the operation, however, and, unfortunately, there was not enough cable to reach shore. The weather was fair and currents low, so it was possible for the deployment vessel to maintain position where the cable ran out, which was approximately 1 to 2 kilometers from shore, and allow a quick initial test of the cabled system to be performed.

Because the cable could not reach shore, the team worked for several hours to terminate and test the cable directly from the vessel. BPPT engineers led by Iyan Turyana spliced connectors onto the fiber to enable communication to the acoustic modem at the cable end (see the series of photos posted to online directory, *Hazardous Seas*, on the Island Press website, https://islandpress.org/tsunami). As soon as the fibers were terminated and the power was connected, the subsea unit was energized with a power supply from the ship. The system came to life after its capacitors charged, demonstrating that the subsea unit at the cable end and the acoustic modem were functional. After the subsea system was up, the team waited for a tense hour until it received the first report transmitted acoustically from the remote instrument 25 kilometers farther away. The signal strength of the acoustic link was good, matching that of the feasibility test performed in 2016 and proving that the acoustic option for connecting remote sensors to the end of a relatively short cable was reliable.

The tidal signal transmitted by the acoustic modem from the remote instrument 25 kilometers away from the cable end using the acoustic modem is shown in figure 8-7. The remote unit was programmed to operate in an hourly mode for the first 60 hours and then revert to once-a-day reporting. Data are also being logged within the instrument for recovery

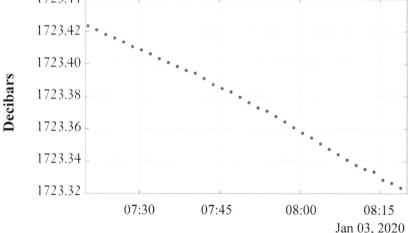

FIGURE 8-7. Tidal signal transmitted by the acoustic modem from the remote instrument 25 kilometers away from the cable end during the short testing period. 1 decibar = 1 meter.

later. Immediate transmission occurs if a signal over a preset threshold is exceeded. The *x*-axis is time, and the *y*-axis is pressure. One decibar is equivalent to 1 meter. Thus, the depth is 1,723 meters.

Photographs, posted on the online directory for this book on the Island Press website, show some of the operations. They also describe different phases of the deployment and the testing done while the cable end was still accessible on the ship, before it was tethered to a buoy and then lowered to the seafloor for recovery and retermination later in 2020.

Conclusion

The real-time seafloor tsunami detection system, coupled with acoustic communications to reach shore via the cabled receiver, represents a new capability in near-shore tsunami detection. The system is uniquely but not solely suitable for areas such as the southern coast of Sumatra, where the detection sensor is in deep water and the water near the surface is sufficiently warm to cause sound rays to bend back toward the seafloor where a cabled receiver can be placed. The initial engineering test proved that the range of the acoustic system would indeed be at least 25 kilometers, providing confidence for the second phase of the program, which was the actual implementation of the detection system for the pressure sensor. The final deployment, although it did not reach its goal of a complete operation system due to the short cable, at least provided a complete end-to-end demonstration of the pressure sensor, acoustic communications system, and cabled acoustic modem receiver.

Future installations are feasible along other areas of both Sumatra and Java, where the water depth close to shore is approximately 1,000 meters within 10 kilometers. The short cable length allows for relatively inexpensive cables to be used and minimizes cost. The end result is another option for policy makers and disaster management professionals to use in constructing a robust approach to near-shore tsunami warning while balancing cost with performance.

References

Bernard, Eddie N., and Christian Meinig. 2011. "History and future of deep-ocean tsunami measurements." In *OCEANS'11 MTS/IEEE KONA*, 1–7. IEEE.

Comfort, L. K., T. Znati, M. Voortman, and L. E. Freitag. 2012. "Early detection of near-field tsunamis using underwater sensor networks." *Science of Tsunami Hazards* 31, no. 4.

Data Buoy Cooperation Panel and International Tsunameter Partnership. 2011. "Ocean data buoy vandalism—incidence, impact and responses," ver. 1. DBCP Technical Document No. 41. World Meteorological Organization and Intergovernmental Oceanographic Commission.

Freitag, Lee, Keenan Ball, Peter Koski, Sandipa Singh, and Eric Gallimore. 2010. "Acoustic communications for deep-ocean observatories: Results of initial testing at the MBARI MARS node." In *OCEANS'10 IEEE SYDNEY*, 1–6. IEEE.

Freitag, Lee, Keenan Ball, James Partan, Peter Koski, and Sandipa Singh. 2015. "Long range acoustic communications and navigation in the Arctic." In *OCEANS 2015-MTS/IEEE Washington*, 1–5. IEEE.

Freitag, Lee, Mark Johnson, Milica Stojanovic, Daniel Nagle, and Josko Catipovic. 2000. "Survey and analysis of underwater acoustic channels for coherent communication in the medium-frequency band." In *OCEANS 2000 MTS/IEEE Conference and Exhibition. Conference Proceedings (Cat. No. 00CH37158)*, vol. 1, 131–38. IEEE.

Freitag, Lee, and Sandipa Singh. 2009. "Performance of micro-modem PSK signaling under variable conditions during the 2008 RACE and SPACE experiments." In *OCEANS 2009*, 1–8. IEEE.

Gallimore, Eric, Jim Partan, Ian Vaughn, Sandipa Singh, Jon Shusta, and Lee Freitag. 2010. "The WHOI micromodem-2: A scalable system for acoustic communications and networking." In *OCEANS 2010 MTS/IEEE SEATTLE*, 1–7. IEEE.

González, Frank I., Eddie N. Bernard, Christian Meinig, Marie C. Eble, Harold O. Mofjeld, and Scott Stalin. 2005. "The NTHMP tsunameter network." *Natural Hazards* 35, no. 1: 25–39.

Heidemann, John, Milica Stojanovic, and Michele Zorzi. 2012. "Underwater sensor networks: Applications, advances and challenges." *Philosophical Transactions of the Royal Society A: Mathematical, Physical and Engineering Sciences* 370, no. 1958: 158–75.

Leonard, Lucinda J., and Jan M. Bednarski. 2015. "The preservation potential of coastal coseismic and tsunami evidence observed following the 2012 Mw 7.8 Haida Gwaii thrust earthquake." *Bulletin of the Seismological Society of America* 105, no. 2B: 1280–89.

Paros, J., P. Migliacio, T. Schaad, W. Chadwick, C. Meinig, M. Spillane, L. Tang, and S. Stalin. 2012. "Nano-resolution technology demonstrates promise for improved local tsunami warnings on the MARS project." In *2012 Oceans-Yeosu*, 1–6. IEEE.

Polet, Jascha, and Hiroo Kanamori. 2000. "Shallow subduction zone earthquakes and their tsunamigenic potential." *Geophysical Journal International* 142, no. 3: 684–702.

Proakis, John G., Ethem M. Sozer, Joseph A. Rice, and Milica Stojanovic. 2001 "Shallow water acoustic networks." *IEEE Communications Magazine* 39, no. 11: 114–19.

Rice, Joe, and Dale Green. 2008. "Underwater acoustic communications and networks for the US Navy's Seaweb Program." In *2008 Second International Conference on Sensor Technologies and Applications (sensorcomm 2008)*, 715–22. IEEE.

Schindelé, F., D. Reymond, E. Gaucher, and E. A. Okal. 1995. "Analysis and automatic processing in near-field of eight 1992–1994 tsunamigenic earthquakes: Improvements towards real-time tsunami warning." *Pure and Applied Geophysics* 144, no. 3: 381–408.

Sheehan, Anne F., Aditya Riadi Gusman, Mohammad Heidarzadeh, and Kenji Satake. 2015. "Array observations of the 2012 Haida Gwaii tsunami using Cascadia Initiative absolute and differential seafloor pressure gauges." *Seismological Research Letters* 86, no. 5: 1278–86.

Stojanovic, Milica, J. A. Catipovic, and J. G. Proakis. 1994. "Phase-coherent digital communications for underwater acoustic channels." *IEEE Journal of Oceanic Engineering* 19, no. 1: 100–111.

Xerandy, X., Taieb Znati, and Louise K. Comfort. "Cost-effective, cognitive undersea network for timely and reliable near-field tsunami warning." *International Journal of Advanced Computer Science and Applications* 6 (2015): 224–33.

Chapter 9

A Prototype Ocean Bottom Pressure Sensor Deployed in the Mentawai Channel, Central Sumatra, Indonesia: Preliminary Results

Emile A. Okal and Lee Freitag

In this chapter, we analyze data retrieved from an ocean floor pressure sensor continuously operated for forty-eight days in Indonesia's Mentawai Channel during the spring of 2016 as part of Project Hazard SEES, Interdisciplinary Research in Hazards and Disasters, funded by the US National Science Foundation. Initial processing through systematic spectrogram analysis has identified eight distant earthquakes recorded through the variation of pressure accompanying the passage of seismic waves on the bottom of the ocean. The analysis of the corresponding wave trains allows the recovery of the standard magnitude M_S of five of the events (two more being intermediate depth and the eighth antipodal) with a residual not exceeding 0.2 logarithmic unit. We also show that the classical energy-to-moment ratio computation can be successfully adapted by defining a response function of the pressure sensor to teleseismic *P* waves. In addition, six local earthquakes, occurring at distances of 58 to 670 kilometers from the sensor, but with magnitudes less than 5.5, were also recorded. We show that an estimate of the seismic energy radiated by these events can be obtained from a simple integration of the square of the pressure signal. Thus, our results indicate that meaningful quantitative estimates of the source characteristics of both teleseismic and regional events can be obtained through robust methods based on single-station pressure recordings on the ocean floor.

This chapter is adapted from L. Freitag and E. A. Okal, 2020, "Preliminary results from a prototype ocean-bottom pressure sensor deployed in the Mentawai Channel, Central Sumatra, Indonesia," *Pure and Applied Geophysics* 177: 5119–31, https://doi.org/10.1007/s00024-020-02561-6.

Background and Motivation: A Layman's Discussion
of the Science behind Seismic Tsunami Warning

This chapter reports on the experimental operation of a prototype ocean bottom pressure sensor in the context of tsunami detection in the Mentawai Channel between Siberut Island and the large island of Sumatra, Indonesia, to the northeast (see the map in the preface).

Sumatra, at the western extremity of Indonesia, borders a so-called subduction zone where the Australian tectonic plate sinks underneath the Eurasian one. This sinking process is taking place principally through large earthquakes, some of them generating catastrophic tsunamis, most recently during the 2004 Sumatra-Andaman event (Synolakis et al. 2005).

Our field location, off the central part of Sumatra, was the site of megaearthquakes in 1797 (northern half) and 1833 (southern half) (Zachariasen et al. 1999; Natawidjdaja et al. 2006). To the south, part of the 1833 fault zone ruptured again during the 2007 Bengkulu earthquake (Borrero et al. 2009), thus releasing at least part of the tectonic stress accumulated over 175 years. By contrast, in the vicinity of Siberut, the plate interface is believed to be locked in the form of a so-called seismic gap, where conditions are now believed ripe for a "megathrust" event to take place in the next years or decades. Such an earthquake would probably generate a tsunami with potentially catastrophic consequences for Padang, a nearby port city with a 2019 metropolitan population of 1.4 million (Borrero et al. 2006).

Tsunami warning rests fundamentally on the detection of the parent earthquake and its real-time interpretation in terms of potential for tsunami generation. In very general terms, all seismic waves can be satisfactorily modeled by representing their source as an appropriate combination of forces (known as a moment tensor) and applying the principles of mechanics to Earth considered as an elastic body. That framework has been extended to the modeling of tsunamis, which can be considered as a particular case of the large family of seismic waves; in simple terms, the mechanical process that excites the waves is the same one that generates the tsunami.

Because of the linearity of the equations of motion, both in elasticity theory and to a large extent in fluid mechanics, one would expect the amplitudes of seismic waves and tsunamis to be scaled and thus for the former to be appropriate predictors of the latter. Such scaling is at the core of tsunami warning. Once the "size" of the earthquake is known, its tsunami potential should be predictable. In this context, and in the near field where warning efforts do not have the luxury of time, a simple message from the scientific community to the populations at risk has consistently been "*the shaking is the warning*," that is, if you feel it and you are close to the shoreline, waste no time and self-evacuate to a safe height, typically 10 meters or more.

In scientific terms, the size of the earthquake has been described since the works of Vvedenskaya (1956) in Russia and Knopoff and Gilbert (1959) in the United States as a bona fide physical measurement, known as the seismic moment M_0 of the system of forces mentioned above, measured in physical units of dynes times centimeters (dyn * cm) or newtons times meters (N * m). Aki (1966) and later Dziewonski et al. (1981) have formalized methodologies for the routine measurements of M_0, now available from worldwide observations within 10 to 20 minutes following an earthquake (Kanamori and Rivera 2008). Kanamori (1977) has derived a protocol to represent M_0 as a so-called moment magnitude M_w designed to be comparable to estimates of conventional magnitudes developed, notably by Richter (1935) and later Gutenberg (1945), at a time when theoretical seismology lacked adequate bases for the representation of seismic sources as moment tensors, but were still widely used in observational seismology and by the media.

In lay terms, the concept of scaling of seismic sources assumes that all their properties (such as seismic slip on the fault plane, length of rupture, areas of given felt intensities, and duration of slip at the source) grow in unison and can be predicted from a single number, namely the seismic moment M_0 or its magnitude rendition M_w. Remarkably, this conjecture has been verified overwhelmingly not only among major earthquakes, but also during minor cracks induced by activity such as mining or fracking and also in laboratory studies on single crystals, across a total of seventeen orders of magnitude for M_0 (Ide and Beroza 2001).

Unfortunately, although most earthquakes are well-behaved and do follow scaling laws, seismological practice has identified in the past few decades a number of anomalous events in clear violation of the paradigm. In very simple terms, one can think of the set of seismic waves generated during an earthquake as a kind of Earth symphony; in most cases, it is perfectly balanced, but some maverick events will emphasize the bass (in spectroscopic terms, they would be called "red-shifted"), while others may favor the trebles (and as such would be "blue-shifted"). In short, for such rogue earthquakes, a single value of seismic moment or moment magnitude may not tell the whole story about their seismic source.[1]

Because tsunamis have periods typically ranging from tens of minutes to 1 hour, their generation is controlled by the lowest-frequency part of the seismic source spectrum, while waves responsible for shaking felt by individuals have typical periods of 1 second or less. In this context, it is clear that anomalous earthquakes featuring an uneven spectrum will violate the expected relationship between the two effects and as such constitute a serious challenge for tsunami warning.

The class of red-shifted, *slow* earthquakes is particularly treacherous, a classic example being the 1992 Nicaraguan tsunami. Its "body-wave" magnitude, measured at a period of 1 second, $m_b = 5.3$, was so low that the earthquake was felt only weakly along certain sections of the shore and even not at all in others. Its source, however, was hiding its full size at longer periods, typically 300 seconds and longer, with a moment magnitude reaching $M_w = 7.6$. The result was a powerful tsunami that reached 10 meters in run-up (defined as the maximum altitude of land inundated by a tsunami) and killed 170 people, eradicating several villages where the earthquake had not been felt (Abe et al. 1993). In such instances, the simple adage "*the shaking is the warning*" cannot apply. In a classic paper based on similar events in 1896 in Sanriku, Japan, and in 1946 in the Aleutian Islands, Kanamori (1972) had coined the name "tsunami earthquakes" for the class of such earthquakes whose tsunamis are much larger than would be predicted by their seismic waves, especially at periods conventionally associated with felt effects.

By contrast, a number of earthquake sources have been found to be "blue-shifted," that is, to have a spectrum enriched in high frequencies, leading to enhanced destruction and casualties through excessive ground accelerations. One such acceleration reached in excess of 20 meters per second squared, or more than twice Earth's gravity, during the 2011 Christchurch, New Zealand, earthquake (Kaiser et al. 2012), an otherwise moderate event ($M_w = 6.2$). Because these sources correspond to a source with a faster than expected stress release, they can be referred to as "brisk" or "snappy."

Motivated by the occurrence of three tsunami earthquakes in Nicaragua in 1992, Java in 1994, and northern Peru in 1996, researchers led a considerable effort to understand the processes leading to departure from scaling laws in the years and decades following these events (Choy and Boatwright 1995; Tanioka et al. 1997; Polet and Kanamori 2000). In particular, Newman and Okal (1998) introduced a slowness parameter Θ, allowing to quantify the anomalous character of an earthquake. Because it requires an estimate of the seismic moment, however, the parameter Θ is generally not immediately available in real time for the benefit of tsunami warning in the near field. In this context, later work by Convers and Newman (2013) and Okal (2013) has focused on exploiting the duration of high-frequency seismic waves as evidence of earthquake slowness.

One of the most fundamental questions regarding tsunami earthquakes is whether they feature a regional component, that is, whether they tend to occur in specific regions or rather could happen along any subduction zone. Many studies, including of historical events predating the development of modern, digital seismic instrumentation, have identified a catalog of more than twenty tsunami earthquakes with documented cases as early as 1896, 1907, and 1923 (Martin et al. 2019, table 6). In this context, preliminary evidence would suggest that many, and probably all, subduction zones can entertain tsunami earthquakes, notably in the form of aftershocks of regular megathrust events (e.g., Fukao 1979; Okal and Borrero 2011; Salaree and Okal 2018).

The departure of seismic sources from the canons of scaling laws took a particularly dramatic turn during the 2009–2010 sequence of events in and around our field area. On September 30, 2009, a catastrophic earthquake hit the city of Padang, causing considerable damage and upwards of eleven hundred deaths (Bothara et al. 2010). However, it generated only a minor tsunami with a maximum run-up of 27 centimeters, due to the earthquake's location at a depth of 80 kilometers, inside the subducting slab rather than at the plate interface, thus offsetting most of the rupture area under Sumatra (in simple terms, the earthquake moved more rock than water). In addition, the 2009 Padang earthquake was clearly of a "snappy," blue-shifted character, with its source not exceeding 10 seconds in duration, later confirmed through a high energy-to-moment parameter Θ (Saloor and Okal 2018). This situation falsely instilled in the minds of residents the idea that significant tsunami danger would require shaking even stronger than during the 2009 earthquake.

Only 13 months later, however, the 2010 earthquake took place seaward of the Mentawai Islands, generating a catastrophic tsunami with a run-up of 17 meters that caused seven hundred deaths on the islands (Hill et al. 2012). This event, which can be construed as an aftershock of the 2007 Bengkulu earthquake to the south, was clearly a "tsunami earthquake," felt at deceptively low levels on the islands, and was later identified as such in a number of seismological studies (Newman et al. 2011; Saloor and Okal 2018). Thus, and tragically, these two events of 2009 and 2010 were both anomalous, but in opposite ways; the 2009 fast event caused enhanced dynamic destruction and death, and the 2010 slow one caused an enhanced tsunami.

Based on reports from survivors who described the 2010 Mentawai event as a "gentle, slow, rocking earthquake that lasted for several minutes" (Hill et al. 2012, 4) and on the work of Convers and Newman (2011), who pointed out its anomalous character within 17 minutes of origin time based on a comparison of the duration and energy of teleseismic *P* waves, this tsunami earthquake would suggest amending the near-field recommendation to "the shaking, *strong or long*, is the warning." The task of issuing a warning, however, runs into the immense difficulty of defining an appropriate level of duration using a qualitative perception of time by lay populations.

In this very general context, a prototype experiment took place in the spring of 2016, when a sensor was deployed for slightly less than two months in the Mentawai Basin. This experiment is the subject of this chapter.

Operational Aspects

Our 2016 experiment consisted of operating a pressure sensor on the bottom of the Mentawai Channel. The technology of the instrument in use is conceptually similar to that of the so-called deep-ocean assessment and reporting of tsunamis (DART) buoys (Meining et al. 2005). Ours, however, is designed to be used in the near field, in the immediate epicentral area. In addition, instead of transmitting via acoustic modem to a nearby surface buoy, our system transmits over distances of 20 to 30 kilometers to a seafloor station cabled to shore. The motivation for such a long-range acoustic link is to eliminate the need for a surface buoy, which requires maintenance and is subject to damage from ocean forces or vandalism (Teng et al. 2010; Mungov et al. 2013).

In addition, we recall that DART buoys operate at a sampling rate of $\delta t = 15$ seconds, that particular channel being triggered for real-time transmission only upon detection of a large event. By contrast, our system uses a much shorter time sampling of $\delta t \approx 0.05$ second, with the real-time transmission of the full dataset made possible by the elimination of the buoy relay. That short time sampling allows a full broadband seismic processing of the time series, in contrast to the coarse sampling used by DART buoys. A permanent seafloor sensor will still require regular maintenance including battery changes, so minimizing energy use is critical. This limitation then mandated the use of a single pressure sensor as opposed to a complete broadband motion package such as those employed in standard ocean bottom seismometers. In this regard, it is desirable to extract as much useful information as possible from the pressure sensor, including the magnitude of seismic waves impinging on the ocean bottom unit. The work presented here reviews the results of an initial approach toward that use.

The instrumental package was deployed in the Mentawai Basin on March 23, 2016, in the vicinity of 1.350°S, 99.733°E at a of depth $H \approx$ 1,750 meters (hereafter Mentawai Basin site or MBS).[2] The above coordinates refer to the sea surface location where the instrument was dropped, which is estimated to coincide with its resting position within uncertainties on the order of a few hundred meters. The exact depth of deployment was provided by the sensor itself. The instrument was retrieved on May 16, 2016, and provided a continuous stream of data from March 26 to May 12, 2016, that is, for forty-eight days.

The instrument deployed consists of a Paroscientific Digiquartz nanoresolution pressure sensor model 8CB2000-I, which includes preprocessing of the native frequency output so that the data are available as an ASCII serial data stream (Paros et al. 2012). The package was deployed on a platform resting on the seafloor and consisted of a pressure sensor, a data logger, a battery, and an acoustic release for recovery.[3] The platform, called a lander because it rests directly on the seafloor, is deployed from a surface vessel and free-falls to the bottom.

Recording was performed at a sampling rate of twenty-two samples per second. The raw data were stored as a pressure time series expressed in pounds[-force] per square inch (psi), later converted to metric units through the factor

$$1 \text{ psi} = 68972 \text{ dyn} \times \text{cm}^2 = 6897.2 \text{ Pa} \tag{1}$$

Figure 9-1 is an example of a 24-hour time series obtained for April 13, 2016, containing the record of an earthquake in Myanmar (hereafter event T4), with an intermediate depth of 136 kilometers and moment magnitude $M_w = 6.9$ (Kanamori 1977). It is easily verified that the average value recorded ($\sim$ 2,550 psi) is the hydrostatic pressure of the water column at the site, $p = \rho_w gH$, which translates to a depth $H = 1,744$ meters using a density $\rho_w = 1.03$ g/cm³ and the acceleration of gravity $g = 979$ cm/s², which is appropriate near the equator. In addition, the large oscillation shown in figure 9-1 is the tidal signal, whose peak-to-peak amplitude, typically 1.15 psi, translates into an amplitude of 79 centimeters for the oceanic tide. This amplitude compares favorably with

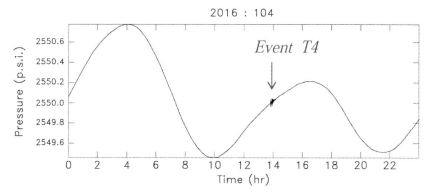

FIGURE 9-1. Example of raw 24-hour time series recorded at the Mentawai Basin site on April 13, 2016. The main oscillation expresses the tide. The signal around 14:00 GMT is Event T4, an intermediate-depth earthquake in Myanmar.

tides typically on the order of 1.2 meters peak to peak in the port of Padang, given the expected influence of the response of the harbor. These observations provide an independent check of the proper calibration of the instrument.

Data Processing

Data processing includes both plotting the raw data from the seismometer and analyzing the spectra to detect seismic events. This process integrates visualization and interpretation of the data to assess the type and size of seismic waves generated by an earthquake with potential tsunami risk and translates the seismic signals into a form that is understood by scientists and practicing emergency managers.

Raw spectra

Figure 9-2 presents the spectrum of a one-day-long window of data, recorded in the absence of detectable seismic signals, on April 9, 2016. In lay terms, it plots the energy recorded by the sensor as background noise, as a function of frequency, from "bass" at the left of the diagram to "treble" at the right. For reference, we compare it to the spectrum of the vertical broadband seismometer at the station GSI, operated by GEOFON at Gunungsitoli, on nearby Nias Island, the distance between the two sites being 380 kilometers.

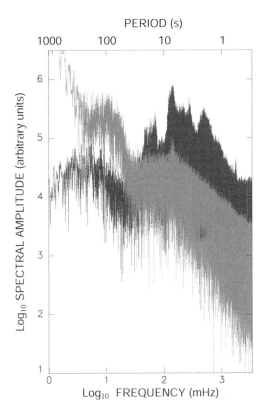

FIGURE 9-2. Spectral amplitude (in grey) of 24-hour record at MBS in the absence of seismic signal (09 April 2016). For reference, the corresponding spectrum at the nearby GEOFON station GSI is given in black. The vertical scales for logarithmic units are common to both plots, but their baselines are different, allowing for direct comparison of the repartition of background noise across the frequency spectrum. Note strong noise at MBS for $T \geq 30s$.

These spectra have not been corrected for instrument response because, in the case of the pressure sensor and as discussed below, the conversion to ground motion involves different functions depending on the nature of the particular seismic wave recorded. Rather, figure 9-2 simply explores the level of background noise, and hence the feasibility of extracting seismic signals in various frequency bands. The vertical scales are common logarithmic units, but unrelated in an absolute sense, thus allowing a *relative* comparison of the levels of background noise as a function of frequency.

Between 0.1 and 10 Hz, corresponding to periods of 10 to 0.1 seconds, the land site GSI is dominated by microseismic noise due to the harmonics of sea swell, peaked around 6 seconds, a ubiquitous feature of coastal seismic stations (Brune and Oliver 1959; Berger et al. 2004; McNamara and Buland 2004). By contrast, and expectedly, these harmonics are absent from the seafloor record, as is the fundamental of the swell, around 12 seconds, thus illustrating the well-known property that submarines do not "feel" the weather, whose relatively short waves fail to penetrate deep into the water column.

Although the noise at the seafloor site is relatively low beyond 30 millihertz, it increases substantially at lower frequencies, in sharp contrast to the seismic spectrum at GSI. This increase will, unfortunately, prevent a quantitative interpretation of surface waves at periods $T \geq 30$ seconds. The origin of this effect is presently unknown.

Spectrogram analysis and detection of seismic events

All forty-eight available 24-hour time windows of data were submitted to a classical spectrogram analysis (Cohen 1989). In simple terms, this procedure isolates a narrow window (in this case, 100 seconds long) moving across the time series (in this case, in steps of 50 seconds) and applies a classic spectral analysis (in our case, between 2 and 10 hertz). The resulting spectral amplitude, computed as in figure 9-2, is then color-coded in decibels with respect to its maximum, with each pixel characterizing the amount of energy present at a given time (abscissa) and frequency (ordinate).[4]

This procedure has been used in various seismological applications for several decades (e.g., Okal and Talandier 1997). Its power is that it allows the systematic detection of small events, which would fail a simple visual investigation, as exemplified by a dynamic profile for the one-day window spanning April 10, 2016: while a small local event, which took place in South Sumatra (hereafter event L2), is clearly visible about 9,000 seconds into the time series, the spectrogram reveals a second earthquake at ~38,000 seconds that would otherwise not emerge from the background noise.

Processing of Teleseismic Events

We call "teleseismic" those earthquakes that occur at a distance of more than 1,000 kilometers from the receiver. Eight such events, listed in table 9-1 and mapped in figure 9-3, were detected on spectrograms in our experiment. Their seismic moments range from 1.1×10^{25} dyn $\times$ cm ($M_w = 6.0$) for event T5, a small earthquake in Mindanao, to 5.9×10^{27} dyn $\times$ cm ($M_w = 7.8$) for event T7, the large 2016 Muisne, Ecuador, earthquake. It is noteworthy that the epicenter of event T7 was essentially antipodal to MBS (the angular distance Δ as seen from the center of the Earth between source and receiver being 178.95° out of a maximum 180°).

TABLE 9-1. Teleseismic events recorded in this study

Date				Origin Time GMT	Epicenter		Depth (km)	Global Values			Distance (°)	Local Estimates			Region
Code	D M (J)		Y		(°N)	(°E)		$M_0{}^a$	M_s	Θ		M_s	E^{Ea}	Θ	
T1	03 APR (094)		2016	08:23:52	−14.32	166.85	26	22	6.8	−4.89	67.51	6.7	35.9	−4.79	Vanuatu
T2	06 APR (097)		2016	14:45:30	−8.20	107.39	29	1.5	5.4	−4.92	10.2	5.4	2.67	−4.75	Java
T3	10 APR (101)		2016	10:28:59	36.47	71.13	212	11		−4.34	46.15		23.5	−4.67	Hindu Kush (Intermediate)
T4	13 APR (104)		2016	13:55:18	23.10	94.68	136	32		−4.46	24.87		114	−4.45	Myanmar (Intermediate)
T5	13 APR (104)		2016	18:21:53	7.79	122.02	17	1.1	5.6	−4.96	24.02	5.7	0.82	−5.13	Mindanao, Philippines
T6	15 APR (106)		2016	16:25:06	32.79	130.75	10	45	7.3	−5.00	44.90	7.2	27.2	−5.22	Kyushu
T7	16 APR (107)		2016	23:58:37	0.35	−79.93	21	590	7.5	−5.15	178.95		167		Ecuador
T8	28 APR (119)		2016	19:33:24	−16.04	167.38	24	36	7.1	−4.71	68.16	6.9		−4.33	Vanuatu

[a]Moment values are in units of 10^{25} dyn × cm, and estimated energies are in 10^{20} erg.

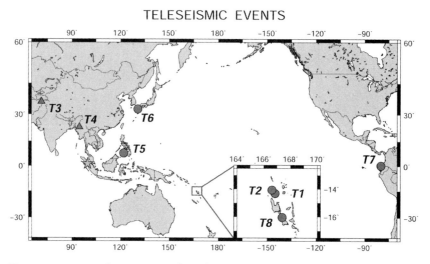

FIGURE 9-3. Map of epicenters of earthquakes detected at teleseismic distances by the hydrophone at the Mentawai Basin site (star). Shallow events ($h < 70$ km) are shown as full circles, and intermediate ones are shown as triangles. See table 9-1 for information on codes T1–T8.

Rayleigh waves

To examine the long-period properties of our records, we first focus on Rayleigh waves, a class of surface waves creeping along the circumference of Earth and prominent at periods of 10 to several hundred seconds (Stein and Wysession 1991). Elementary seismic theory (e.g., Haskell 1953) shows that a pressure sensor at the interface between a solid half-space and an ocean of thickness H records a Rayleigh wave as an over-pressure of amplitude

$$P = \rho_w \omega^2 H \cdot u_z \qquad (2)$$

where u_z is the vertical seismic displacement of the solid Earth and $\omega = 2\pi/T$ is the angular frequency of the wave of period T. In other words, the pressure sensor functions as an accelerometer whose gain is proportional to the depth of the water column. At a depth of 1,750 meters, the vertical displacement of the Rayleigh wave can be restored by first converting the digital values (pounds-force per square inch) to metric units (dyn/cm^2) and then representing the sensor as an instrument featuring two null "zeros" and no poles (Aki and Richards 2002, 637), with a total magnification of $-\rho_w H = -1,802,000$ kilograms per square meter. In simple terms, this procedure allows us to transform the signal detected by the sensor (in units of pressure) into a quantitative measurement of the amplitude of the Rayleigh wave (in units of length).

Because of the excessive noise at periods longer than 30 seconds (figure 9-2), it was not possible to compute mantle magnitudes (Okal and Talandier 1989), which would have allowed the retrieval of a long-period seismic moment at periods of several hundred seconds. Rather, we had to limit our investigations to conventional surface-wave magnitudes, using the Prague formula for Rayleigh waves with a period T close to 20 seconds (Vaněk et al. 1962):

$$M_s = \log_{10}\frac{(A)}{(T)} + 1.66 \log_{10}\Delta + 3.3 \qquad (3)$$

where A is in microns, T in seconds, and Δ in degrees. For each record, we extract a 1-hour time window containing the Rayleigh wave train and filter it between 10 and 30 seconds. A time-domain measurement of the maximum amplitude can be converted into a displacement using equation (2), which substituted into equation (3) leads to

$$M_s = \log_{10}(p \cdot T) + 1.66 \log_{10}\Delta + 5.3 \qquad (4)$$

where p is in psi. Figure 9-4 shows that the records of teleseismic events are comparable to standard seismograms and illustrates the measurement of M_s according to equation (4). As detailed in table 9-1, we obtain M_s values in excellent agreement with published ones; note that we do not compute M_s for events T3 and T4, which are at intermediate depths where 20-second surface waves are poorly excited, nor for T7, for which the station is antipodal ($\Delta \approx 179°$).

We conclude that the pressure sensor can be used reliably to quantify the amplitude of conventional 20-second Rayleigh waves from teleseismic events in the range $M_s = 5.4$ to 7.2. Unfortunately, the only significantly larger event recorded during our deployment was antipodal; also, the presence of unexplained but substantial noise at longer periods prevented the extension of our investigations to the domain of mantle waves more directly associated with tsunami excitation.

P waves

P (for "primary") waves are seismic body waves, traveling from a source to a receiver through the Earth's interior ("body") (Stein and Wysession

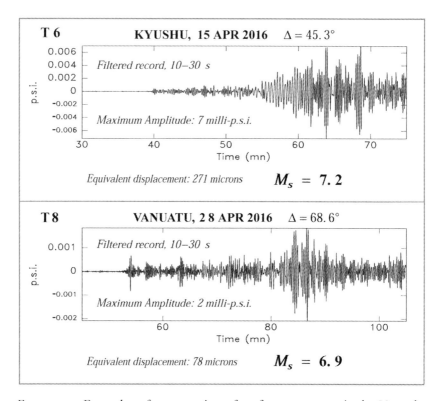

FIGURE 9-4. Examples of computation of surface-wave magnitude M_s at the Mentawai Basin site for teleseismic events T6 (Kyushu) and T8 (Vanuatu). See text for details.

1991). They constitute the fastest signals reaching a distant station, in practice a few minutes after origin time. They are also least attenuated during their propagation and as such carry significant energy even at short periods, typically on the order of 1 second.

Here, we use P waves recorded by our ocean bottom pressure sensor from teleseismic events to compute an energy flux at the receiver, and from there an estimate of the seismic energy radiated by the source, following the algorithm of Newman and Okal (1998), itself inspired by Boatwright and Choy (1986). Such measurements constitute a quantification of the earthquake at the high-frequency end of its source spectrum.

At the bottom of a liquid layer where the sound velocity is α_w, it can be shown that, upon incidence of a P wave, the ratio of pressure in the fluid to vertical displacement at the interface (known in physics as an impedance) is

$$Z = \frac{P}{u_z} = \rho_w \alpha_w \omega \qquad (5)$$

This formula shows that a pressure sensor responds differently to Rayleigh waves and P waves. Whereas in the former case it behaved as an accelerometer, it will now respond to ground velocity, which amounts to adding a factor $(\alpha_w/\omega H)$ to the response used above for surface waves.

The computation of radiated energy proceeds along the steps detailed by Newman and Okal (1998) and routinely implemented since then, but an additional complexity stems from multiple reverberations in the water column (figure 9-5). As an incident seismic wave impinges on the surface of the Earth, it undergoes significant transformations to prevent its continuation into the atmosphere (conveniently taken as a vacuum), which can be regarded as a relatively complex extension of the principle of an optical mirror. As a consequence, the field of ground motion is altered at the free surface, requiring a correction in the computation of the teleseismic energy flux, noted $C^p(i_0)$ in Newman and Okal (1998), and whose detailed expression is given, for example, by Okal (1992). In the oceanic environment, the water layer traps a fraction of the energy in a frequency-dependent process reminiscent of the effect of coating in the conception of partial mirrors. It can be considered the conjugate, at the receiver, of the well-known source-side generation of multiply reflected $pmwP$ phases (Mendiguren 1971). At the high frequencies characteristic of P waves, it is appropriate to sum the energies of the various rays involved in

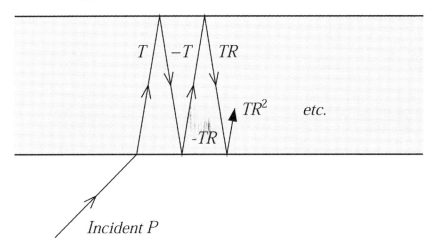

FIGURE 9-5. Multiple reflections and response function of the oceanic column for a teleseismic P wave recorded as a pressure signal at the bottom of the ocean. See text for details.

the multiple reflections (see figure 9-5), and the classical surface response coefficient for incident P waves, used by Newman and Okal (1998), must be replaced by a more complex one, whose detailed expression is derived in Okal and Freitag (2020). Finally, a special algorithm is used for events T3 and T4, whose depths are intermediate (Saloor and Okal 2018), and an additional correction effected for events T2, T4, and T5, for which the station is less than 30° away (Ebeling and Okal 2012).

Table 9-1 includes values of the resulting estimated energies E^E for seven teleseismic events (the computation is not carried out for the antipodal event T7), as well as parameters $\Theta = \log_{10}(E^E/M_0)$, obtained using published values of the seismic moments M_0 of the relevant earthquakes. These values are compared to values of Θ computed routinely from a global dataset of stations (Newman and Okal 1998; Saloor and Okal 2018). As shown in figure 9-6, the agreement is excellent, with no systematic trend in the residual between the value of Θ obtained here from the pressure sensor and its reference value; as for the root-mean-square of the residual (0.22 logarithmic unit), it is comparable to the scatter of individual station values when using large global datasets at seismological stations. Our results thus validate the use of the pressure sensor to

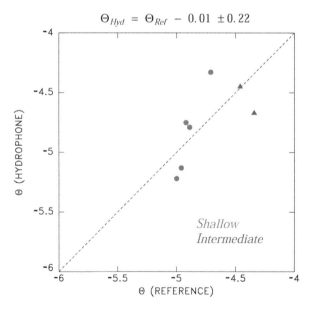

FIGURE 9-6. Comparison of values of the slowness parameters Θ obtained from a global dataset (abscissa) and from the hydrophone at the Mentawai Basin site (ordinate).

quantify a teleseismic source at the higher-frequency end of the seismic spectrum.

Processing of Local Events

Six regional earthquakes, listed in table 9-2 and mapped in figure 9-7, were detected from spectrograms in the present experiment. Two of them, events L1 and L5, took place at intermediate depths of 160 and 115 kilometers, respectively, in the down-going slab under Sumatra. Except for the small event L6, also closest to the sensor at only 58 kilometers, all had a moment tensor inverted as part of the Global Centroid-Moment-Tensor, or Global CMT, Project (Dziewonski et al. 1981; Ekström et al. 2012), with moments listed in table 9-2.

Also shown in figure 9-7 are "beachball" diagrams of the relevant moment tensors, which offer a conventional representation of the geometry of faulting of the earthquake source (Stein and Wysession 1991). Note the diversity in focal geometries, which illustrates the oblique convergence at the Sumatra trench (Sella et al. 2002).

At such regional distances, the formalism of Newman and Okal (1998) no longer applies, and an estimate of earthquake energy must be obtained through an alternate computation. A simple approach consists of directly integrating the energy flux of the time series of overpressure. By analogy with energy estimates computed at teleseismic distances (Boatwright and Choy 1986; Newman and Okal 1998), we simply consider the pressure flux

$$F_p = \frac{\alpha_w}{\pi} \int \frac{|P|^2(\omega)}{K} \cdot d\omega \qquad (6)$$

where $P(\omega)$ is the Fourier transform of the pressure $p(t)$ and where α_ω and K are the sound velocity and bulk modulus of the water, respectively; again by analogy with computations of estimated energy, the integral in equation (6) is conveniently limited to the window 0.1 to 2 hertz. Note that we neglect anelastic attenuation at regional distances. Through a further, and admittedly drastic simplification, an estimate of the total energy of the seismic source is then obtained by scaling F_p to the square of the epicentral distance D and weighting the result to include the contribution of S waves to obtain a pressure-estimated energy

$$E_p^E = 4\pi \cdot D^2 \cdot (1 + q^{BC}) \cdot F_p \qquad (7)$$

TABLE 9-2. Local events recorded in this study

Code	Date D M (J) Y	Origin Time GMT	Epicenter (°N)	Epicenter (°E)	Depth (km)	Global Values M_0^a	Global Values m_b	Distance (km)	Local Estimates F_p^a	Local Estimates E_p^{Ea}	Local Estimates Θ_p	Region
L1	29 MAR (089) 2016	06:24:48	−2.807	102.319	160	8.4	5.1	329.9	24.6	0.56	−5.18	Southern Sumatra (Intermediate)
L2	10 APR (101) 2016	02:14:35	−4.149	102.211	41	46	6.0	414.9	194.	6.99	−4.57	Southern Sumatra
L3	15 APR (106) 2016	10:24:31	−3.620	100.492	38	5.8	4.7	265.3	12.4	0.21	−5.47	Southern Sumatra
L4	16 APR (107) 2016	21:09:12	0.450	97.971	10	2.1	5.1	279.7	6.84	0.11	−5.25	Northern Sumatra
L5	02 MAY (123) 2016	04:21:25	−4.989	104.551	115	47	6.0	669.8	32.7	3.06	−5.19	Southern Sumatra (Intermediate)
L6	03 MAY (124) 2016	22:32:36	−1.875	99.734	35	2.9[b]	4.5	58.2	3.30	0.20	−5.16[b]	Southern Sumatra

[a] Moment values are in in units of 10^{23} dyn × cm, fluxes are in g/s^2, and estimated energies are in 10^{19} erg.

[b] For Event L6, values in italics are estimated from scaling laws (Geller 1976; Okal 2019).

LOCAL EVENTS

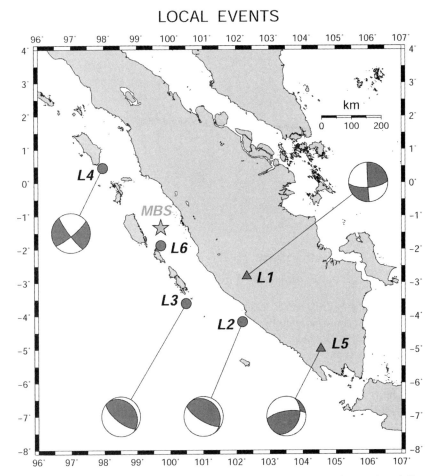

FIGURE 9-7. Map of the epicenters of local earthquakes detected at regional distances by the hydrophone at the Mentawai Basin site (star). Shallow events ($h <$ 70 km) are shown as full circles, and intermediate ones are shown as triangles. See table 9-2 for information on codes L1–L6. Also shown are global CMT solutions, except for event L6.

where q^{BC} is given by Newman and Okal (1998) after Boatwright and Choy (1986). For event L6, at a distance smaller than 100 kilometers, the far-field approximation to the body wavefield inherent in those authors' formalism breaks down (Aki and Richards 2002, section 4.3), and D is replaced in the near field by $D^{NF} = 2\pi D^2/\Lambda$, where the wavelength $\Lambda =$ 40 kilometers is computed at a typical period of 5 seconds.

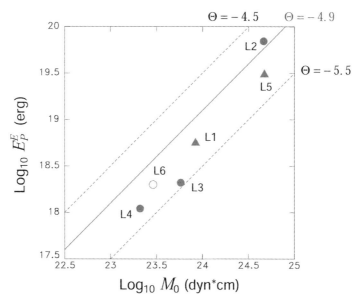

FIGURE 9-8. Pressure-estimated seismic energy E_p^E at MBS *vs.* published seismic moment M_0. Circles refer to shallow events, triangles to intermediate ones. The open circle is event L6, whose moment is estimated on the basis of its body-wave magnitude m_b.

The results are listed in table 9-2, where E_p^E is scaled to the published M_0, yielding a pressure-estimated energy-to-moment parameter Θ_p. Figure 9-8 plots E_p^E versus published moment M_0.

It is not possible to compare directly Θ_p to values of Θ computed from seismograms at teleseismic distances because the regional events detected during our deployment are too small ($M_0 \leq 4.7 \times 10^{23}$ dyn * cm; $M_w \leq 5.7$) to lend themselves to a routine global computation of Θ. Nevertheless, it is remarkable that this range of computed values of Θ_p (−4.57 to −5.47) is typical of values measured at subduction zones (Okal and Newman 1998). In the case of the smallest (and closest) event L6, for which no seismic moment is available, a tentative value of $M_0 = 2.9 \times 10^{23}$ dyn * cm ($M_w = 4.9$) can be estimated by scaling the moment of event L3 using the difference of body wave magnitudes (4.5 versus 4.7), in a range of sizes where m_b has started to saturate (Geller 1976; Okal 2019), and under the assumption that those events follow scaling laws. The result is a tentative value of $\Theta_p = -5.18$, again in excellent agreement with values expected at subduction zones, confirming that an estimate of high-frequency quantification of seismic sources in the near field can be obtained, even at very short distances and for small events.

In summary, this experiment with pressure records of regional earthquakes shows that it is feasible to obtain quantitative estimates of the size of seismic sources. In practice, we did not catch any obviously anomalous regional earthquakes during our short-lived experiment. We emphasize, however, that the events detected were all small by global seismological standards, and, *a fortiori*, had no tsunamigenic potential.

Conclusion

During a forty-eight-day window of continuous operations, we were able to detect fourteen earthquakes—six at regional distances and eight global events at teleseismic distances—for which meaningful quantifications of the seismic source were obtained and, for all but one small event, successfully compared to globally published values. These results verify the concept of using a seafloor pressure sensor to quantify the source of a seismic event, particularly in the near field, even though the presence of unexpected noise at periods longer than 30 seconds restricts the recovery of the longest-period part of the seismic spectrum.

There remains the caveat that, given the short nature of our experiment, neither a truly large earthquake (except the antipodal Ecuadorian event T7), nor a significant local one, was detected at MBS. In particular, none of the fourteen events generated any gravitational water wave detectable by our pressure sensor. Only the continuous operation of the sensor over a much longer period of time will allow us to eventually expand our investigation to this condition, in fulfillment of the motivation of our project.

Acknowledgments

This research was supported by the Hazards SEES program of the National Science Foundation under Grant Number OCE-1331463 to the University of Pittsburgh. We thank Louise Comfort for her leadership in this project. The authors would like to acknowledge the invaluable help of Professor Febrin Anas Ismail of Andalas University (Padang), Iyan Turyana of the BPPT (Jakarta), and the captain and crew of the *KN MUCI* (Padang). The Woods Hole Oceanographic Institution field team included engineers Peter Koski and Keenan Ball with additional support from Sandipa Singh and James Partan. Some figures were plotted using the GMT software (Wessel and Smith 1991).

Notes

1. Incidentally, a very similar situation exists regarding hurricanes, commonly classified into "categories" (ranging from 1 to 5) supposedly describing all their properties (size of the system, maximum wind velocity, underpressure at the center of the eye). Recent cases (e.g., Hurricanes Sandy in 2012 and Patricia in 2015) have shown that these properties are not always correlated and that a classification using a single category number constitutes a significant, and potentially hazardous, oversimplification.

2. A map showing the location of the Mentawai Basin site can be found in the Resources tab at islandpress.org/hazardous-seas.

3. A photo of this undersea unit is available in the Resources tab at islandpress.org/hazardous-seas.

4. A color-coded dynamic profile for the one-day window of April 10, 2016, can be found in the Resources tab at islandpress.org/hazardous-seas.

References

Abe, Kuniaki, Katsuyuki Abe, Yoshinobu Tsuji, Fumihiko Imamura, H. Katao, I. Yohihisa, Kenji Satake, Joanne Bourgeois, E. Noguera, and F. Estrada. 1993. "Field survey of the Nicaragua earthquake and tsunami of September 2, 1992." *Bulletin of the Earthquake Research Institute, University of Tokyo* 68, no. 1: 23–70.

Aki, Keiiti. 1966. "Generation and Propagation of G Waves from the Niigata Earthquake of June 16, 1964.: Part 2. Estimation of earthquake moment, released energy, and stress-strain drop from the G wave spectrum." *Bulletin of the Earthquake Research Institute, University of Tokyo* 44, no. 1: 73–88.

Aki, Keiiti, and Paul G. Richards. 2002. *Quantitative Seismology*, 2nd ed. Sausalito, CA: Univ. Science Books, 218–35.

Berger, Jonathan, Peter Davis, and Göran Ekström. 2004. "Ambient Earth noise: A survey of the global seismographic network." *Journal of Geophysical Research: Solid Earth* 109, no. B11: B11307, 10 pp.

Boatwright, John, and George L. Choy. 1986. "Teleseismic estimates of the energy radiated by shallow earthquakes." *Journal of Geophysical Research: Solid Earth* 91, no. B2: 2095–112.

Borrero, José C., Kerry Sieh, Mohamed Chlieh, and Costas E. Synolakis. 2006. "Tsunami inundation modeling for western Sumatra." *Proceedings of the National Academy of Sciences* 103, no. 52: 19673–77.

Borrero, José C., Robert Weiss, Emile A. Okal, Rahman Hidayat, Suranto, Diego Arcas, and Vasily V. Titov. 2009. "The tsunami of 2007 September 12, Bengkulu province, Sumatra, Indonesia: Post-tsunami field survey and numerical modelling." *Geophysical Journal International* 178, no. 1: 180–94.

Bothara, Jitendra, Dick Beetham, Dave Brunsdon, Mike Stannard, Roger Brown, Clark Hyland, Warren Lewis, Scott Miller, Rebecca Sanders, and Yakso Sulistio. 2010. "General observations of effects of the 30th September 2009 Padang earthquake, Indonesia." *Bulletin of the New Zealand Society for Earthquake Engineering* 43, no. 3: 143–73.

Brune, James N., and Jack Oliver. 1959. "The seismic noise of the Earth's surface." *Bulletin of the Seismological Society of America* 49, no. 4: 349–53.

Choy, George L., and John L. Boatwright. 1995. "Global patterns of radiated seismic energy and apparent stress." *Journal of Geophysical Research: Solid Earth* 100, no. B9: 18205–28.

Cohen, Leon. 1989. "Time-frequency distributions—a review." *Proceedings of the Institute of Electrical Electronic Engineering* 77, no. 7: 941–81.

Convers, Jaime Andres, and Andrew V. Newman. 2013. "Rapid earthquake rupture duration estimates from teleseismic energy rates, with application to real-time warning." *Geophysical Research Letters* 40, no. 22: 5844–48.

Dziewonski, Adam M., T.-A. Chou, and John H. Woodhouse. 1981. "Determination of earthquake source parameters from waveform data for studies of global and regional seismicity." *Journal of Geophysical Research: Solid Earth* 86, no. B4: 2825–52.

Ebeling, Carl W., and Emile A. Okal. 2012. "An extension of the E/M_0 tsunami earthquake discriminant Θ to regional distances." *Geophysical Journal International* 190, no. 3: 1640–56.

Ekström, Göran, Meredith Nettles, and Adam M. Dziewoński. 2012. "The Global CMT Project 2004–2010: Centroid-moment tensors for 13,017 earthquakes." *Physics of the Earth and Planetary Interiors* 200: 1–9.

Freitag, L., and E. A. Okal. 2020. "Preliminary results from a prototype ocean-bottom pressure sensor deployed in the Mentawai Channel, Central Sumatra, Indonesia." *Pure and Applied Geophysics* 177: 5119–31. https://doi.org/10.1007/s00024-020-02561-6.

Fukao, Yoshio. 1979. "Tsunami earthquakes and subduction processes near deep-sea trenches." *Journal of Geophysical Research: Solid Earth* 84, no. B5: 2303–14.

Geller, Robert J. 1976. "Scaling relations for earthquake source parameters and magnitudes." *Bulletin of the Seismological Society of America* 66, no. 5: 1501–23.

Gutenberg, Beno. 1945. "Magnitude determination for deep-focus earthquakes." *Bulletin of the Seismological Society of America* 35, no. 3: 117–30.

Haskell, N. A. 1953. "The dispersion of surface waves on multilayered media." *Bulletin of the Seismological Society of America* 43: 17–34.

Hill, Emma M., José C. Borrero, Zhenhua Huang, Qiang Qiu, Paramesh Banerjee, Danny H. Natawidjaja, Pedro Elosegui, et al. 2012. "The 2010 M_w = 7.8 Mentawai earthquake: Very shallow source of a rare tsunami earthquake determined from tsunami field survey and near-field GPS data." *Journal of Geophysical Research: Solid Earth* 117, no. B6: B06402, 21 pp.

Ide, Satoshi, and Gregory C. Beroza. 2001. "Does apparent stress vary with earthquake size?" *Geophysical Research Letters* 28, no. 17: 3349–52.

Kaiser, A., C. Holden, J. Beavan, D. Beetham, R. Benites, A. Celentano, D. Collett, et al. 2012. "The M_w = 6.2 Christchurch earthquake of February 2011: Preliminary report." *New Zealand Journal of Geology and Geophysics* 55, no. 1: 67–90.

Kanamori, Hiroo. 1972. "Mechanism of tsunami earthquakes." *Physics of the Earth and Planetary Interiors* 6, no. 5: 346–59.

Kanamori, Hiroo. 1977. "The energy release in great earthquakes." *Journal of Geophysical Research* 82, no. 20: 2981–87.

Kanamori, Hiroo, and Luis Rivera. 2008. "Source inversion of W phase: Speeding up seismic tsunami warning." *Geophysical Journal International* 175, no. 1: 222–38.

Knopoff, Leon, and Freeman Gilbert. 1959. "Radiation from a strike-slip fault." *Bulletin of the Seismological Society of America* 49, no. 2: 163–78.

Martin, Stacey S., Linlin Li, Emile A. Okal, Julie Morin, Alexander E. G. Tetteroo, Adam D. Switzer, and Kerry E. Sieh. 2019. "Reassessment of the 1907 Sumatra 'tsunami earthquake' based on macroseismic, seismological, and tsunami observations and modeling." *Pure and Applied Geophysics* 176, no. 7: 2831–68.

McNamara, Daniel E., and Raymond P. Buland. 2004. "Ambient noise levels in the continental United States." *Bulletin of the Seismological Society of America* 94, no. 4: 1517–27.

Meinig, Christian, Scott E. Stalin, Alex I. Nakamura, Frank González, and Hugh B. Milburn. 2005. "Technology developments in real-time tsunami measuring, monitoring and forecasting." In *Proceedings of OCEANS 2005 MTS/IEEE*, pp. 1673–79. IEEE.

Mendiguren, Jorge A. 1971. "Focal mechanism of a shock in the middle of the Nazca plate." *Journal of Geophysical Research* 76, no. 17: 3861–79.

Mungov, George, Marie Eblé, and Richard Bouchard. 2013. "DART® tsunameter retrospective and real-time data: A reflection on 10 years of processing in support of tsunami research and operations." *Pure and Applied Geophysics* 170, no. 9–10: 1369–84.

Natawidjaja, Danny H., Kerry Sieh, Mohamed Chlieh, John Galetzka, Bambang W. Suwargadi, Hai Cheng, R. Lawrence Edwards, Jean-Philippe Avouac, and Steven N. Ward. 2006. "Source parameters of the great Sumatran megathrust earthquakes of 1797 and 1833 inferred from coral microatolls." *Journal of Geophysical Research: Solid Earth* 111, no. B6: B06403, 37 pp.

Newman, Andrew V., Gavin Hayes, Yong Wei, and Jaime Convers. 2011. "The 25 October 2010 Mentawai tsunami earthquake, from real-time discriminants, finite-fault rupture, and tsunami excitation." *Geophysical Research Letters* 38, no. 5: L05302, 7 pp.

Newman, Andrew V., and Emile A. Okal. 1998. "Teleseismic estimates of radiated seismic energy: The E/M_0 discriminant for tsunami earthquakes." *Journal of Geophysical Research: Solid Earth* 103, no. B11: 26885–98.

Okal, E. A. 1992. Use of the mantle magnitude M_m for the reassessment of the moment of historical earthquakes." *Pure and Applied Geophysics (PAGEOPH)* 139: 17–57. https://doi.org/10.1007/BF00876825.

Okal, Emile A. 2013. "From 3-Hz P waves to $_0S_2$: No evidence of a slow component to the source of the 2011 Tohoku earthquake." *Pure and Applied Geophysics* 170, no. 6–8: 963–73.

Okal, Emile A. 2019. "Energy and magnitude: a historical perspective." *Pure and Applied Geophysics* 176, no. 9: 3815–49.

Okal, Emile A., and José C. Borrero. 2011. "The 'tsunami earthquake' of 1932 June 22 in Manzanillo, Mexico: Seismological study and tsunami simulations." *Geophysical Journal International* 187, no. 3: 1443–59.

Okal, Emile A., and Jacques Talandier. 1989. "M_m: A variable-period mantle magnitude." *Journal of Geophysical Research: Solid Earth* 94, no. B4: 4169–93.

Okal, Emile A., and Jacques Talandier. 1997. "T waves from the great 1994 Bolivian deep earthquake in relation to channeling of S wave energy up the slab." *Journal of Geophysical Research: Solid Earth* 102, no. B12: 27421–37.

Paros, J., P. Migliacio, T. Schaad, W. Chadwick, C. Meinig, M. Spillane, L. Tang, and S. Stalin. 2012. "Nano-resolution technology demonstrates promise for improved local tsunami warnings on the MARS project." In *2012 Oceans-Yeosu*, pp. 1–6. IEEE.

Polet, Jascha, and Hiroo Kanamori. 2000. "Shallow subduction zone earthquakes and their tsunamigenic potential." *Geophysical Journal International* 142, no. 3: 684–702.

Richter, Charles F. 1935. "An instrumental earthquake magnitude scale." *Bulletin of the Seismological Society of America* 25, no. 1: 1–32.

Salaree, Amir, and Emile A. Okal. 2018. "The 'tsunami earthquake' of 13 April 1923 in Northern Kamchatka: Seismological and hydrodynamic investigations." *Pure and Applied Geophysics* 175, no. 4: 1257–85.

Saloor, Nooshin, and Emile A. Okal. 2018. "Extension of the energy-to-moment parameter Θ to intermediate and deep earthquakes." *Physics of the Earth and Planetary Interiors* 274: 37–48.

Sella, Giovanni F., Timothy H. Dixon, and Ailin Mao. 2002. "REVEL: A model for recent plate velocities from space geodesy." *Journal of Geophysical Research: Solid Earth* 107, no. B4: ETG-11, 32 pp.

Stein, Seth, and Michael Wysession. 1991. *An Introduction to Seismology, Earthquakes, and Earth Structure.* Blackwell.

Synolakis, Costas, Emile Okal, and Eddie Bernard. 2005. "The mega-tsunami of December 26, 2004." *Bridge*, 35, no. 2: 26–35.

Tanioka, Yuichiro, Larry Ruff, and Kenji Satake. 1997. "What controls the lateral variation of large earthquake occurrence along the Japan Trench?" *Island Arc*, 6, no. 3: 261–66.

Teng, Chung-Chu, Stephen Cucullu, Shannon McArthur, Craig Kohler, Bill Burnett, and Landry Bernard. 2010. *Buoy Vandalism Experienced by NOAA National Data Buoy Center.* National Oceanic and Atmospheric Administration Stennis Space Center MS National Data Buoy Center, 10 pp.

Vaněk, Jiří, Alois Zátopek, Vit Kárník, Natalya V. Kondorskaya, Yuriĭ V. Riznichenko, Evgeniĭ F. Savarenskiĭ, Sergeĭ L. Solov'ev, and Nikolaĭ V. Shebalin, 1962. Standardizatsya shkaly magnitud, *Izv. Akad. Nauk SSSR, Ser. Geofiz.*, 2, 153–158, 1962 [in Russian]; English translation, *Bull. USSR Acad. Sci.*, 2:108–11.

Vvedenskaya, A. V. 1956. "Opredelenie poleí smeshcheni pri zemletryaseniyakh s pomoshchyu teori dislokatsiĭ." *Izv. Akad. Nauk SSSR, Ser. Geofiz.* 3: 277–84.

Wessel, Paul, and Walter H. F. Smith. 1991. "Free software helps map and display data." *EOS, Transactions American Geophysical Union* 72, no. 41: 441, 445–46.

Zachariasen, Judith, Kerry Sieh, Frederick W. Taylor, R. Lawrence Edwards, and Wahyoe S. Hantoro. 1999. "Submergence and uplift associated with the giant 1833 Sumatran subduction earthquake: Evidence from coral microatolls." *Journal of Geophysical Research: Solid Earth* 104, no. B1: 895–919.

Chapter 10

Underwater Sensor Network Prototype for Tsunami Detection and Warning: A Long Deployment Journey toward Functionality

X. Xerandy, Iyan Turyana, Lee Freitag, Wahyu W. Pandoe,
Harkunti P. Rahayu, and Febrin Anas Ismail

The last critical step in this research is to demonstrate that the prototype system is working and feasible in an actual environment. We will not be bringing the system into an operational state, however. As mentioned in previous chapters, the deployment takes place in the West Sumatra region Indonesia.

The implementation of the prototype involves international collaboration between the University of Pittsburgh, the Woods Hole Oceanographic Institute (WHOI), the Indonesian Agency for the Assessment and Application of Technology (Badan Pengkajian dan Penerapan Teknologi, BPPT), the Bandung Institute of Technology, Bandung (Institute Teknologi Bandung, ITB), and Andalas University, supported by grants from the Swiss Re Foundation. The deployment effort began in December 2020.

This chapter presents the process and steps taken to implement the component systems of the prototype. This account provides insight into the technical complexity and challenges that the research team addressed during the implementation of the system components at the project site. Furthermore, nontechnical challenges such as the COVID-19 outbreak and the drastic organizational changes for BPPT contributed substantially to the demands of system implementation. Figure 10-1a illustrates the general architecture of the system prototype, and figure 10-1b shows the location for the deployment. A detailed description of the whole underwater system is presented in chapter 8.

The prototype implementation consists of four main field tasks: (1) cable-laying operation, (2) canister fabrication and deployment, (3) undersea node's battery replacement and renovation, and (4) landing station development and satellite connection. Because the prototype benefits

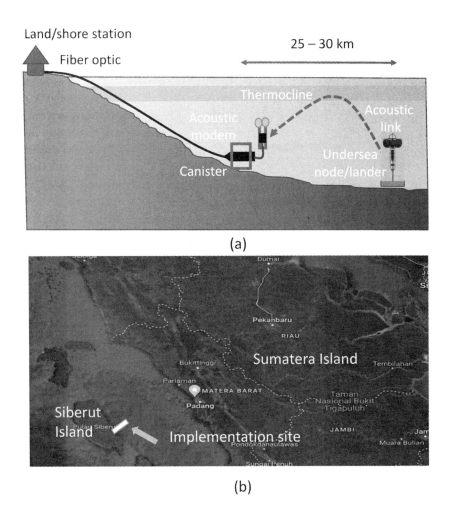

Land/shore station

Fiber optic

25 – 30 km

Thermocline

Acoustic modem

Acoustic link

Canister

Undersea node/lander

(a)

Dumai

Pekanbaru

RIAU

Sumatera Island

Tembilahan

Bukittinggi

Pariaman

SUMATERA BARAT

Padang

Taman Nasional Bukit Tigapuluh

Siberut Island

Pulau Siberut

Implementation site

Pondokdanaulawas

JAMBI

Jam

Muara Bulian

Sungai Penuh

(b)

Figure 10-1. (a) Implementation process and (b) site for prototype deployment. (Diagram by Xerandy.)

from the presence of the thermocline layer, the canister deployment must be deeper than this layer. In Indonesia, this layer typically exists at the depth between 150 to 400 meters below the ocean surface. The seafloor at the implementation site near Siberut Island features extreme slopes and rugged terrain. To deploy the prototype in such a harsh undersea environment imposed unexpected demands in both technical and funding aspects. The cost incurred for the implementation was considerably higher than estimated for the scale of this research. The submarine fiber-optic cable and deployment operation accounted for the largest portion of the cost in this deployment.

In common practice, a ship equipped with a dynamic positioning (DP) system is required. A DP system is an advanced navigation function that enables a ship to maintain a steady position and heading in rolling ocean waters, controlled by a system of integrated propellers and thrusters. For this operation, the ship also needed on-deck equipment such as a cable engine, winch, and A-frame gantry crane to support the submarine cable–laying work. To minimize the cost, some compromises on cable-laying practices and techniques were made. Working within budget constraints, this approach was unavoidable, despite increasing the risk of failure.

As the result, the deployment has taken a long journey toward completion. Three cable-laying missions and one mission of undersea node maintenance operation have been undertaken since January 2020 to implement the prototype. The first two undersea cable-laying missions encountered critical points of failure, but despite the difficulties, they represent important steps in the discovery of technical requirements for operating in a deep-ocean environment. They became learning opportunities to bring the third mission to success.

The First Deployment: November 2019 – January 3, 2020

The initial effort began on November 11, 2019, by negotiating a memorandum of understanding between BPPT, WHOI, and the University of Pittsburgh that was signed in San Francisco. A memorandum of understanding is required to provide the legal basis for research activities that would be conducted in Indonesia. The first deployment was conducted from the last week of December 2019 through the first week of January 2020. The tasks in the first deployment consisted of laying the cable, canister fabrication, and landing station development.

Fiber-optic cable-laying operation and canister fabrication

In this collaboration, BPPT was responsible for assembling the canister and, together with the cable-laying company, assisted in setting up the cable-laying procedure. BPPT was also responsible for building the landing station. To reduce the cost, the team chose a local cable-laying company with little experience working in deep water.

In normal submarine cable–laying practices, a detailed mapping of the seafloor where the cable would be laid is done to determine the cable-laying route. This mapping is typically executed by using a multibeam sonar technique. For this project, however, the sea mapping was made using a single-beam technique that was executed repeatedly to obtain relatively dense data. The mapping activity was made long before this implementation mission and was done in conjunction with a different project. The mapping result was not as detailed as that using multibeam sonar, but it was considered sufficient to determine the route for a short cable-laying operation. Figure 10-2 shows roughly the seafloor terrain where the cable would be laid.

The plan was to lay the cable from the sea to the shore. As figure 10-2 suggests, the laying route stretches along a seafloor valley. The underwater seafloor contour poses extreme terrain features and steep slopes, making the laying operation more challenging. The sea bottom with this contour likely has coral reef and hard rocks. Therefore, stronger protection to the submarine cable should be applied. In standard submarine cable–laying practices, simple and lightweight cables are used because they account for the longest portion of the cable to be deployed. The lightweight cable, however, lacks sufficient protection against a harsh environment as presented near Siberut Island. Extra steel armor is usually added to give extra protection to the submarine cable, but this method considerably increases the cable cost.

Despite the harsh sea bottom, this project used a lightweight type of submarine cable to stay within funding limits, with the expectation that the cable could withstand this harsh environment for one or two years, a much shorter period than the commercial underwater submarine cable. The deployment did not use a DP-equipped vessel; instead, a smaller ship with a double propeller system was considered sufficient for this operation. Having two propellers allows a vessel to be more agile, especially during maneuvering at low speed. The project secured a vessel of this type to perform the deployment, the MV *Bakat Menuang* (figure 10-3). This vessel is a small ro-ro (roll on–roll off) vessel used to transport vehicles, goods, and people between small islands in West Sumatra. Further, a simple, inexpensive design for the cable engine was used and installed on the ship, with the assumption that the workload on the engine would not be high because the cable length was relatively short.

The result did not come as expected. Because the ship did not have good braking equipment, cable payout measurement, or tension monitoring equipment, the deployment of the 7.3 kilometer submarine fiber-optic cable ended about 2 kilometers short of the shore. This limited equipment resulted in the uncontrollable release of the cable from the ship. The problem was further exacerbated by not managing the vessel's

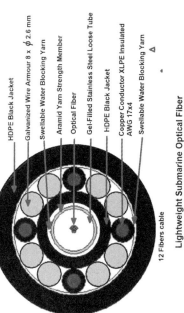

HDPE Black Jacket

Galvanized Wire Armour 8 x φ 2.6 mm

Swellable Water Blocking Yarn

Aramid Yarn Strength Member

Optical Fiber

Gel-Filled Stainless Steel Loose Tube

HDPE Black Jacket

Copper Conductor XLPE Insulated AWG 17x4

Swellable Water Blocking Yarn

12 Fibers cable

Lightweight Submarine Optical Fiber

(b)

Acoustic Link

Optical Fiber cable

(a)

FIGURE 10-2. (a) The submarine cable route and (b) the lightweight cable cross-section used in this project. (Diagram by Xerandy and Iyan Turyana.)

FIGURE 10-3. The vessel used for the first deployment, MV *Bakat Menuang*. (Photo from BPPT archives.)

speed to compensate for the rapid cable payout due to inconsistent communication and coordination between the ship's deck crew and the ship's bridge staff who were responsible for navigation. This experience underscored the need for an experienced company with professional equipment to lay the cable in such a harsh environment.

The Second Deployment: July 2020

Because the otherwise successful deployment of January 3, 2020, did not achieve the final termination of the fiber-optic cable on shore, a second deployment was scheduled, initially for March 2020 but delayed by the outbreak of COVID-19 until July 2020. During this period, several improvements were made to the landing station on shore in preparation for a more robust operation of the undersea acoustic network, explained below.

Landing station development

The submarine fiber-optic cable terminates at a landing station on shore. This project's landing station is basically a building or platform that houses the power and communication devices needed to run the prototype. It is located approximately 150 meters inland from the coastline and behind the police station in the town of Muara Siberut and was built by engineers from Andalas University, Padang, West Sumatra, in cooperation with ITB.

The landing station provides a direct current (DC) power supply to the canister and acoustic modem. The power supply gets energy from either a solar power grid or a battery. The battery is recharged by a solar cell system affixed to the roof of the landing station. The data transmitted from the canister over the fiber-optic cable will be forwarded to the central station in Jakarta using either a satellite or cellular network, whichever is available.

Figure 10-4a shows the working progress of excavating a small channel toward the landing station in which the submarine cable will be buried, and the weatherproof box in which the power supply system is installed is shown in figure 10-4b. Figure 10-4c shows the terminations of optical fibers in the landing station, which carry the information from the undersea node. More specifically, the terminations are attached to an optical-to-serial converter device.

Learning from mistakes made in the first mission, the second mission was rescheduled for late March 2020, then delayed further due to the

(a)   (b)   (c)

FIGURE 10-4. Landing station. The development progress showing (a) solar panels, (b) power system and optical to serial devices, and (c) the receiver of optical transmission from the cable. (Photos courtesy of Xerandy.)

emergence of COVID-19, which caused nations to close their borders to international travel. Given this unforeseen disruption, the Swiss Re Foundation offered a supplementary grant to cover costs for COVID-19 testing and quarantine in early June 2020. After carefully following protocols for minimizing risk of exposure to engineers and crew, the second mission was rescheduled for mid-July 2020, to be carried out by the Indonesian engineers with the WHOI engineers communicating with the team via virtual electronic media.

In the second mission, the Swiss Re Foundation provided funding for the purchase of a new cable and canister and the cost of a professional submarine cable–laying operator. The existing submarine cable deployed in the first mission could not be reused and had to be pulled up from the sea bottom. To support the second mission, BPPT contributed the use of the cable-laying ship RV *Baruna Jaya III* (figure 10-5) for ten days, including crew and fuel. The ship, owned by BPPT, is approximately 60 meters in length, with approximately 1,000 gross tons of weight.

The ship does not have a DP system. To compensate for the lack of this equipment on the ship, we used a more sophisticated cable-laying engine and hired additional professional cable-laying technicians. Further, the cable-laying procedure was thoroughly studied and discussed together with cable-laying professionals, especially because we were using a non-DP vessel. There were two critical activities in the operation: recovering the existing canister and acoustic receiver and laying the new cable and canister. Communication and coordination between personnel on board were also evaluated and improved. To ensure the success of the cable-laying operation, RV *Baruna Jaya III* conducted a maneuvering simulation to follow the given laying track. To increase laying safety during maneuvering, detailed bathymetric measurements were taken, particularly in the near-shore area.

A new canister was also manufactured to replace the existing canister. The procurement of the new canister was managed by ITB, and the design and manufacturing of the canister housing was done by the same company, CCSI, that manufactured the submarine fiber-optic cable. The electronic system was designed and implemented by BPPT. There was no significant change in the electronic circuit in the canister. BPPT also provided the canister frame for this second mission.

Initially planned to commence in late March 2020, the second mission was rescheduled amid the COVID-19 pandemic, making its completion more challenging. Mobility was strongly restricted. Not only were the WHOI engineers not allowed to travel due to this pandemic, but BPPT

FIGURE 10-5. The vessel used for the second deployment, RV *Baruna Jaya III.* (Photo from BPPT archives.)

also imposed a strict policy to mitigate the spread of infection. Dispatching the research vessel required all members of the ship's onboard crew to pass a COVID-19 screening test. Once the crew members passed this test, they had to isolate or quarantine for at least three days in a designated place before departure. No persons were allowed to enter the ship except the crew. This policy incurred extra costs. The team submitted a supplemental budget to the Swiss Re Foundation to cover the extra costs incurred by the pandemic. As the result, these requirements pushed the deployment schedule back to July 2020, or three months behind the initial planned date. The second deployment was conducted from July 19 to August 1, 2020.

The second cable-laying mission met its goal. The existing canister and the modem were successfully recovered, and the cable reached the shore. This operation was not completed smoothly, however: when doing the onboard communication testing, the onboard team faced an unexpected and unanticipated electronic problem when the canister could not communicate with the modem. To overcome this problem, the onboard electronic engineer needed to splice the cable. The recommended tool and materials to do the work on board were not available, so splicing was conducted using very limited equipment. Although this solution was far from ideal, it would solve the problem if were done properly. Having that problem in the field, especially given that the biggest technical challenge is recovering the canister, a light, simple mooring line attached to the canister and extending to the surface was proposed. This construction allows for easier recovery of the canister in the future.

Although the cable-laying operation was considered successful, the electronic system was not. When the system was powered up, excessive power consumption was detected. Based on the initial assessment, the probable cause of the problem was water ingression into the canister. There was no way to validate the problem without pulling back the canister, however, and doing so at that moment was impossible. Thus, the deployment was considered partially successful in the second mission. The research team decided to launch a third attempt to complete the deployment in 2021.

Regarding the continuing threat of COVID-19, the team learned an important lesson during this deployment: most members of the ship's crew involved in the mission became infected by COVID-19. The ship's health and safety executive on board suspected that this situation might be caused by inconsistent obedience of the ship's crew in complying with the strict onboard isolation policy. Due to the unexpected technical problem during deployment, however, the onboard team had no choice but to allow local people to board the ship to assist with the cable-laying operation. This decision violated the onboard isolation policy, but it was unavoidable. Fortunately, all crew members survived the disease.

The Third Deployment: Final Attempt, November–December 2021

Learning from the previous two failures, more serious deployment planning was conducted for the third deployment, which was planned as the final attempt. We strove to design, prepare, and use all tools and methods that are technically essential to ensure a successful deployment. In this phase, the paradigm was that cost must follow the technical need, not vice versa. It is also notable that all preparations and execution of the third deployment were taken just as the Omicron variant of COVID-19 surged during the pandemic, making it very challenging to accomplish.

We learned two lessons from failure in the second deployment, mostly due to having loose control during the undersea node-making process. First, the process of canister manufacturing must be made under tight control and monitoring. The canister must pass pressure testing for deployment in deep water. In this context, BPPT committed to provide a new canister for this final deployment and would supervise the design, manufacture, and testing of a new canister to ensure that no leakage occurs after deployment. BPPT contracted with a third-party company to design and manufacture the canister. Second, the electronic system must be designed more carefully. Technical coordination between the BPPT and WHOI teams in designing the canister's electronic system was also intensified to prevent communication failure. A functional test was also designed to ensure that the whole system works properly.

Considering the complexity of this deployment, a DP-equipped vessel was considered essential to provide sufficient maneuverability and stability to carry out the delicate operation of recovering the canister and replacing it at sea. Through a collaborative arrangement between the Swiss Re Foundation, the University of Pittsburgh, and ITB, the remaining funds in the research grant from the Swiss Re Foundation administered by the University of Pittsburgh were reallocated to ITB to provide a DP-system vessel for deploying the prototype.

The team's further concern for the third deployment was the working status of the stand-alone undersea sensor node, called a lander (see figure 10-1a). It was deployed on January 3, 2020, and had been sitting on the seafloor since that time. This device is powered by an internal battery that has a limited lifetime, by design, of about two years, after which the lander or ocean bottom unit needs to be recovered for maintenance. Leaving the device longer increases the risk of device loss due to recovery failure. Engineers from WHOI expected to do the recovery on the same mission as replacing the canister in this final deployment in December 2021. Although international travel from the United States to Indonesia had been easing, visa provision was suddenly closed again in late November due to the surge of the Omicron variant in the COVID-19 pandemic. Without visas, the WHOI engineers could not travel to Indonesia. This restriction canceled the battery maintenance task, and no battery replacement or ocean bottom unit maintenance could be carried out during this third deployment.

Finally, the research team not only had to manage the technical complexity of the engineering requirements and global pandemic, but also had to face drastic organizational change at BPPT when the agency was disbanded on September 1, 2021. This action impacted key agency decision makers who had supported the project. Consequently, many internal chains of command and administrative procedures needed to be adjusted following the change. Most of BPPT's main programs were reevaluated. The change began in the third quarter of 2021, when the procurement of the canister was in progress. This situation slowed the process of canister procurement and fabrication, which created an unavoidable delay for the research project. Nevertheless, the research team members from BPPT succeeded in securing the procurement and fabrication of the canister by the last week of October 2021.

Supervising the design of the canister and canister frame

Because a suspected problem in the previous mission was imperfect canister production, supervision of the design and fabrication of the canister and canister frame became a critical part of the third mission's preparation. BPPT took serious steps in preparing the canister and contracted with a submarine fiber-optic cable and equipment factory to manufacture the canister. A series of technical requirements were imposed during production of the canister. The canister should be strong enough to withstand the pressure caused by the seawater at the depth of the deployment. Furthermore, functional tests on the communication and electronic system were conducted at every critical point of the canister manufacturing process.

The canister was made slightly larger than the previous version to accommodate a larger space for electronic system assembly and integration. The outer cap of the canister was also redesigned to give more space for external connector plugging activity. The canister frame, the platform to hold the canister, was also designed carefully. This aspect had not been carefully supervised in the previous two missions. The canister frame was designed to be heavier to hold the mooring line. The frame was arranged so that a technician is still able to access the canister to plug the cable to canister connector. The frame was made from corrosion-resistant stainless steel. The canister holder locking mechanism is designed to anticipate axial force during deployment. Although the frame was made from corrosion-resistant stainless steel, a carbon steel plate was also installed on the frame as sacrificing anode. This measure adds extra protection to the frame from corrosion.

Supervising the pressure test of the canister

The canister will be deployed on the seafloor at the depth of 900 meters under water pressure equal to 90 bar or metric units of water pressure.

The canister needs to withstand such pressure. Following good design practice, the canister is designed to withstand at least 1.5 times of the required pressure, or about 135 metric units. Thus, we designed the canister to hold under pressure up to 150 bar. Special attention was given to the sealing mechanism in the canister endcap.

BPPT required that the manufacturer of the canister conduct strict pressure tests to ensure that the canister housing is watertight. The test was done not once, but three times. Between tests, the canister endcap should be dismantled and reassembled again. This step is taken to ensure that the disassemble-assemble activity on the canister endcap will not compromise the strength of the canister against required pressure, which is especially important during electronic system integration into the canister. The integration was made before the final pressure test. The pressure test was conducted at the company facility selected by BPPT to fabricate the canister. BPPT engineers watched and supervised every test session to ensure that the test was carried out properly. All parts of the canister housing were inspected at the end of each test session to ensure that no water intrusion mark was detected in the endcap sealing.

The canister passed all pressure test sessions. No leakage or water ingression was observed. This result added confidence to the research team that the canister was safe to deploy at the current depth of 700 meters. Figure 10-6a shows the canister pressure test to ensure that the canister can withstand water pressure up to 150 bar, or equal to a water depth of 1,500 meters. Figure 10-6b shows the tension stress test to canister coupling to ensure that the coupling can withstand a pulling force up to two metric tons.

Designing, assembly, and integration of the canister electronic system

The other critical part of this mission is designing the electronic system within the canister properly. BPPT was also responsible for this work. The engineers from the Indonesian Technology Center for Disaster Risk Reduction Technology (Pusat Teknologi Reduksi Risiko Bencana) and the Center of Marine Survey Technology (Balai Teknologi Survey Kelautan) were appointed to do the work. Some modifications were made to the electronic system to enhance the performance of the system. The landing station supply voltage to the system was increased to prevent further loss on power transmission over the fiber-optic cable. This change required all components to be recalculated carefully, so new components were procured to accommodate the change. High-pass filters were also

(a)   (b)   (c)

Figure 10-6. (a) Test of canister to ensure that it can withstand water pressure to a depth of 1,500 m; (b) tension stress test to canister coupling to ensure that it can withstand a pulling force up to 2 tons; (c) canister installed in its frame. (Photos courtesy of Xerandy.)

incorporated to the power regulator to extend the component's lifetime. Further, to utilize the spare conductor line in the fiber-optic cable, a test line was proposed. This test line will reroute the communication line in the canister back to landing station, creating a communication loop line. This method is useful for communication line checking, especially to determine which part of the line is broken. If a communication loop is formed by triggering the test line, the fiber-optic cable and communication component are still in a working state.

Functional tests on the power system and communication were also intensified. The electronic system must undergo and pass these tests at every predetermined critical phase during canister manufacturing and deployment: at assembly; during integration, pressure testing, loading, and transportation; and immediately before being launched to the sea. Those critical phases may potentially cause damage to the canister. Thus, having more testing on the canister ensures the system's integrity and functionality. Figure 10-6c shows the canister installed in its frame, ready for deployment.

Deployment preparation

The research team started to set up the deployment operation plan more carefully after the midyear 2020 deployment. The target date for the deployment was set for the last week of November 2021. For the operation, the research team hired the same cable-laying professional as employed in the second mission. During the canister replacement operation plan review, three technical issues were raised: obtaining the proper ship, prechecking the landing station, and locating the marker buoy of the mooring line that was installed in the second deployment. These aspects are important to guarantee that all serious efforts made through the canister production phases would be reinforced. Indeed, those issues can entail a considerable increase of expenditures for the research team, particularly the rental of a DP-equipped ship. The team needed to find additional sources of funding for a DP-equipped ship, landing station inspection, and locating the marker buoy, as acknowledged in the preface.

As described before, a simple mooring line that is attached to the cable bottom was set up in the second deployment to allow the team to pull up the canister more easily when necessary. We designed the mooring line to have a primary floater just about 15 meters under the sea surface to keep the line straight up with positive buoyancy. This depth is relatively safe from collision by nearby fishing ships but is still reachable by a diver. Then, a light mooring line was extended beyond the floater's depth toward the surface to hold the surface buoy. Having the buoy on the surface helps mark the location where the canister is placed but may attract nearby fishing ships and increase the risk of vandalism. Therefore, the surface buoy was designed to have the least-noticeable shape by reducing its size.

Because the system had been left for almost six months without regular inspection, the research team was concerned about the condition of the mooring line and if it had been broken or collapsed. A small team of cable-laying technicians was assigned to inspect the mooring line to determine if it still existed and do some repairs if necessary to keep it intact until the upcoming canister replacement operation date. The cable-laying technician located the marker buoy and checked the mooring line's condition. The marker buoy's coordinates for its actual location had shifted away about 1 mile (roughly 1.6 kilometers) from the designated coordinates.

The other task was to inspect the landing station. The system had been left unpowered; thus, no active system was working in the landing station. The team wanted to check the latest condition of the landing station and investigate whether damage had occurred to the system, especially to the cable. The solar-generated power supply was also checked for proper functionality. A BPPT technical team, consisting of five engineers, departed to Muara Siberut to assess the latest condition of the landing station. The inspection showed that the battery-bank charging system was still working properly. A different result, however, was found during cable inspection when the team was only able to ascertain that two of four conductors were intact; the other two showed signs of a possible short-circuit, but the team could not determine precisely which sections of the system's subsea cable had short-circuits.

Getting the ship to fit the task

The most difficult task was to find a DP-equipped vessel for the mission. A dedicated deep-sea cable-laying ship is equipped with an advanced DP system. The number of ships of this type is limited in Indonesia, however, and most had been exclusively reserved for Indonesian telecommunication companies to do submarine fiber-optic cable repair. The cost to rent such a ship is enormously expensive. Therefore, a general-purpose DP-equipped working vessel could be an option for this mission, which comes with a less expensive rental rate. Our choice went to an anchor, handling, tug, and supply (AHTS) type of vessel, which is commonly utilized by oil companies to do maintenance work near their offshore oil platforms. An AHTS is usually equipped with an onboard crane and winch, important onboard equipment for the canister handling and replacement operation.

ITB was responsible for finding such vessel, but getting a vessel for a third deployment was not easy in terms of minimum days of use, availability, and the initial embarkation port of the ship. All the ships that the team identified were fully booked until the end of October 2021. The nearest embarkation port of the ships was on Batam Island, which is just across the Singapore Strait from Singapore. It takes about seven days of sailing from Batam to reach our deployment site in Muara Siberut, West Sumatra. The team expected that the actual work for canister replacement would be only three to four days at most. Another complication was the required minimum days of use. The ship's owner required us to rent the ship for at least fifteen days. According to our plan, we only expected to spend eleven days at sea for this work, where most of the time was spent for transit purpose. The team did not see a strong reason to pay for extra ship days that were not needed.

Finally, with some luck, ITB found a vessel with a competitive rental rate and availability that fit our schedule, the MV *Surf Mandiri* (figure 10-7), an AHTS type of vessel. It was idle between two reservations for cable work, and its embarkation port was Jakarta. The vessel was equipped with a DP2 system, with two DP systems installed in the vessel, which was

FIGURE 10-7. The vessel used for the third deployment, MV *Surf Mandiri*. (Photo from BPPT archives.)

technically ideal for our mission. The MV *Surf Mandiri* has a large deck area at the stern of the ship, giving the cable-laying professional more space for working area and putting more equipment. A winch and crane are also available on this ship.

Deployment operation

The deployment operation was conducted from December 2 to December 12, 2021. The BPPT team departed for Padang City, West Sumatra, the closest major city to deployment site, where they met with the canister manufacturer and cable-laying professionals and held a final technical meeting before the operation.

The deployment team was organized into two teams, the sea team and the shore team. The sea team—BPPT electrical engineers, a submarine cable jointer, a canister manufacturer representative, and the cable-laying professionals—would stay onboard the MV *Surf Mandiri* during deployment. The shore team—BPPT electrical engineers and fiber-optic technicians—would stay onshore to perform submarine cable checks and tests at the landing station during canister replacement.

Equipment loading. Once the reservation of the vessel was settled, the canister replacement operation commenced. The MV *Surf Mandiri* was scheduled to pick up the canister in Padang City, so the canister manufacturer was required to transport the canister and submarine cable jointer from its manufacturing plant in Cilegon, Banten, to Padang. Just before being packaged for transport, the BPPT team performed a last functional test on the canister to ensure its electronic system functionality.

Retrieval of old canister and installation of new canister. The crucial tasks of the operation are the retrieval of the old canister and installation of the new canister. During the operation, the sea team determined the submarine conductor cable polarity, checking and performing fiber-optic cable splicing and monitoring the submarine cable jointing operation so that no further damage occurred. The cable-laying professional did the canister recovery from the sea bottom. This work came with the risk of an unexpected cable twist, however, because having the canister sitting on the seafloor at the depth around 900 meters and having no information about how the cable lay on the seafloor increased the chance of twist even more. The risk is that the fiber-optic cable may be damaged and unusable. The twisted section of the cable must be removed, hence making the entire length of the existing cable shorter. If recovery is not done carefully and properly, multiple cable twists might happen and the cable would have

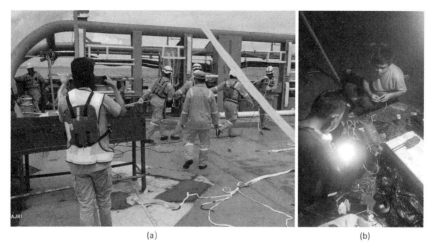

(a)   (b)

FIGURE 10-8. (a) Deck activity during old canister recovery. (b) Shore team checking the cable during retrieval operation of the old canister. (Photos courtesy of Xerandy.)

to be cut repeatedly. The worst-case situation that might occur would be having the entire fiber-optic cable being cut, which would be an unexpected and unwelcome outcome. Figure 10-8a shows the deployment activity on the MV *Surf Mandiri*'s deck when recovering the old canister.

The shore team stayed inland at the landing station location and were available to assist the onboard team (sea team and cable-laying professionals) during the conductor polarity checking process. The sea team also checked for communication functionality during the deployment of the new canister.

Meanwhile, to ensure the integrity of the fiber-optic cable, the fiber-optic technician member of the shore team performed an optical time-domain reflectometer (OTDR) test to monitor optical loss on the fiber and ensure that the loss was still within the required specification. This test was very important, especially during the process of lifting the existing canister up to the surface, when the chances of cable twist were high. A cable twist would lead to a high optical loss. The OTDR test checked for optical loss on the fiber and measured the distance of abnormal event occurrence from where the measurement was taken. Figure 10-8b shows the activity of the shore team performing this test.

The recovery of the canister took longer than expected. Although the cable-laying professionals had done their best to lift the old canister, the cable unexpectedly kinked in multiple points along the cable. By conducting the OTDR test, it was shown that only 4,300 meters of fiber-optic cable were usable. This shortened cable length left the submerged end at a sea depth of 700 meters.

Because underwater acoustic communication between the stand-alone underwater sensor node, referred to as the lander, and the acoustic receiver modem uses the thermocline layer of seawater, having too shallow depth where the canister and acoustic receiver modem were deployed might deteriorate the quality of the acoustic link. This layer typically exists at a depth of approximately 300 meters below the sea surface and is generally found in an open, tropical sea area. As the expert for underwater communication and the institution that developed the acoustic modem for this system, the research team member from WHOI considered that the seawater environment at a depth of 700 meters was still feasible for the acoustic modem to receive sound traffic from the lander node.

The Fourth Mission: Maintenance Operation, April 2022

Because the lander could not be retrieved from the seafloor for maintenance, particularly for battery replacement during the third deployment in December 2021, the research team still needed to conduct the battery replacement operation and landing station check on the shore.

Before being disbanded, BPPT had the responsibility of providing an appropriate ship to conduct the battery replacement and lander maintenance and had secured the funds to cover the cost of ship rental in 2021. Given the delays incurred by COVID-19 travel restrictions, however, the execution took place the following year, in April 2022. The persistent constraints on securing visas that had blocked travel to Indonesia for the WHOI engineers were finally resolved after the Indonesian government relaxed its national border policy to allow international travel. Maintenance supplies such as the replacement battery and toolkit for acoustic release also cleared the Indonesian customs process.

About three months after the completion of the third deployment, the research team planned to perform the landing station checks again. This inspection was necessary to ensure that the cable and canister were still intact and functioning after the third deployment. BPPT engineers undertook this task and also provided the new circuit design of the DC power supply system. The system still used a combination of solar system and power grid as the source of power. The voltage to the system was increased from 200 VDC to 360 VDC to improve its efficiency; also, an automatic current discharging circuitry was added to improve reliability.

Getting the ship ready for the task

Because the lander maintenance task was not conducted during the third deployment, BPPT had funds from its 2021 allocated budget to rent another ship to conduct maintenance. Because this task was less complex than replacing the canister, the ship did not have to be equipped with a DP system and heavy-duty crane. Instead, a smaller ship with a simple lifting tool was needed to lift the lander, which would float to the sea surface after being released.

After careful searching, BPPT found a suitable ship, KM *Eva03*, for this task at a cost within its allocated budget. This local cargo ship that transports goods to nearby islands was based in Padang. The owner of the ship was helpful and did not charge a high rental rate because he realized that the purpose of this mission was to increase the Siberut community's awareness of potential tsunami risk.

The KM *Eva03*, which was initially reserved to do the work in 2021, was rescheduled to April 2022 to fit the schedule of engineers from WHOI, who finally managed to get visas for travel to Indonesia. The maintenance task required two cruises by the ship, which was not initially the case, but the owner of the ship offered to support the second cruise at no extra cost if the team could arrange the second cruise date to coincide with the KM *Eva03*'s scheduled cargo delivery time. BPPT and the ship's owner thus agreed to do the work in the second week of April 2022.

The first trip retrieved the lander from the sea bottom, using the acoustic modem release. After the lander was recovered and loaded onto the ship, the ship returned to Padang carrying the lander. With support from Professor Febrin Ismail at Andalas University, the lander was taken to the university, where laboratory space and workspace were made available in the civil engineering department and WHOI engineers could conduct the necessary maintenance, including battery replacement, firmware update, and download of recorded data from the lander's data disk.

Battery replacement and undersea node maintenance

In April 2022, Indonesian visa requirements for visitors from the United States changed, and it was possible for the team from WHOI to travel to Indonesia to help with the recovery of the lander system to change its batteries and then redeploy it. Lee Freitag and Dennis Giaya of WHOI flew to Jakarta, where they met with Iyan Turyana and Xerandy on April 15 to coordinate the activities, including two short cruises: one for recovery and one for redeployment. The process was completed in just over one week.

The lander consists of two separate systems, each with its own battery. The pressure sensor and detector subsystem operates at very low power and is on continuously; the acoustic modem, however, is normally off except when it transmits the daily report or an event detected on the pressure sensor. The two units operate from data packs built from lithium-ion primary cells, but they are different sizes to account for the differing amounts of energy that are needed. The battery packs were procured from a vendor in the United States and shipped to Indonesia in advance of the instrument turnaround.

The equipment needed for the turnaround was moderate and included hand tools, test cables, and an acoustic deck box that sends a coded sonar signal to the lander to release the weight that is keeping it on the seafloor. That latter device was hand-carried from WHOI to Indonesia. The weight itself was made from iron plates that were fabricated in a local machine shop, which was arranged by Iyan Turyana. A special bit that holds the weights to the acoustic release was already in Jakarta at the BPPT facility. There is also a battery in the acoustic release, bringing the total number of systems that required opening and battery change to three.

After the initial meeting in Jakarta, the team traveled to Padang to start the work. The materials and tools needed for the turnaround effort were staged at the civil engineering department laboratory and offices prior to departure. The recovery cruise left from the river port in Padang and utilized a small wooden freighter that was chartered to motor to the lander position. The ship departed in the evening of April 16 and arrived at the location at 6 a.m. the following day.

The seas were very calm, and the recovery of the lander system was straightforward. The acoustic deck box was used to wake up the acoustic release and measure the travel time to provide the range. It was determined that the ship was directly over the lander as expected, so the command to drop the weight was sent and an acknowledgment received. It took approximately 30 minutes for the unit to come to the surface, and the brightly colored floats were quickly spotted by the crew. The ship approached slowly until the lander and the modem were immediately alongside. A member of the crew then jumped into the water to attach a line to the lander, and the lander was lifted out of the water and onto the

(a)   (b)

FIGURE 10-9. (a) Lander recovery. (b) Lander frame before cleaning. (Photos courtesy of Lee Freitag.)

ship using the ship's boom and a motorized winch to pull the line. See figure 10-9. The process was repeated for the modem system before the ship turned and headed back to Padang, where it arrived in the late afternoon.

Initial inspection of the two systems showed considerable corrosion on both instrument frames, but the titanium instrument housings, including the pressure sensor (made in the United States by Paroscientific), were pristine. During the transit back to port, the pressure sensor and the electronics housing with its battery and the detection processor were removed from the lander frame so that they could be carried to Andalas University. The modem instrument frame was unbolted from the float frame above it but kept in its own frame to make it easier to transport. The acoustic release that drops the weight was also removed from the frame so that its battery could be changed.

On the morning of April 18, the group arrived at Andalas University to begin work on the systems. The pressure sensor instrument housing was opened first, and the data from the SD card was downloaded. It was found that the instrument had operated for more than a year, storing data and the daily updates that were sent to the acoustic modem for regular transmission to the other acoustic modem at the end of the cable. The fresh battery was installed, although there was an issue with the battery pack as wired by the vendor, requiring rewiring to provide the proper voltage. The batteries for both the acoustic modem and the acoustic release were also changed, without any significant issues.

The log files on the SD card, along with detection of several seismic events, showed that the system had been functional, so the decision was made to not change any software to avoid introducing new issues. There was not sufficient time to test any changes, even if they were minor, so the system was simply restarted with a newly formatted data card. There were some issues reading the card on the system that was recovered, but the majority of the data was able to be downloaded.

The aluminum instrument frames for both the lander and the acoustic modem had suffered considerable corrosion, as shown in figure 10-10a, although much of it was cosmetic; in some areas, several millimeters had been lost. The lander had not been anodized, but simply painted with a marine-grade coating. The lander had already been deployed for nearly a year during the previous experiment, and some pitting had occurred then, but considerably more damage occurred during 2020 and 2021. The amount of material loss was not enough to jeopardize structural integrity because of the relatively low loads, however, and the lander was deemed good enough to withstand another deployment of one to two years. A new one will have to be fabricated for any subsequent deployments.

The work continued through April 19 and 20, and on April 21 the systems were loaded back onto the freighter for the deployment cruise. Whereas the first trip had been with the ship empty, this time it was loaded with supplies for Siberut Island, as the second trip was expected to coincide with the ship's cargo delivery schedule. The lander and modem instrument frame were placed on top of rebar, and gas tanks and wheelbarrows were on the forward part of the deck. The final cleanup of the instrument frames occurred on board the ship on April 21, and the fresh anchor was loaded under the lander so that it was ready for deployment, as shown in figure 10-10b. After some delays attributed to bureaucratic

(a)   (b)

FIGURE 10-10. (a) Lander after cleaning. (b) Lander ready for deployment. (Photos courtesy of Lee Freitag.)

issues that required the group to stay one more night in Padang, the ship left in the afternoon on April 22 for Siberut and the work site.

The deployment site was adjusted slightly toward shore to account for the end of the cable being moved from the very first deployment in January 2020. In addition, because of the potential for blocking by an undersea ridge, the location was shifted to try to ensure that there was a clear acoustic path from the lander to the modem at the end of the cable.

The ship arrived at the desired location after dark. The deployment proceeded amid moderate waves from a 15 to 20 knot wind without incident and was completed in less than 15 minutes. The order of the work was to deploy the acoustic modem first, and after it drifted away, the lander was picked up and put over the side. With the lander in the water and falling, the ship backed away, and the modem instrument frame was pulled under very soon. The ship then proceeded to Siberut, where the shore team, including Xerandy, had arranged transportation to the town where the cable comes ashore.

Landing station checks after lander redeployment

Although the team left the new canister and landing station properly connected after the third deployment, the research team thought that rechecking the landing station and submarine cable integrity was urgently necessary, considering the rugged sea bottom terrain and the nearly four-month time lag between the third deployment and battery maintenance of the lander. This consideration was also motivated by the research team's experience during recovery of the former canister in the third deployment, where the team found that some of power lines and optic fibers in the submarine cable had broken. Then, only two of four power lines were available for use, leaving the team with no backup line.

Landing station checks and the lander maintenance were set to be conducted in the same trip, which took place during the second week of April 2022. BPPT engineers were responsible for examining the landing stations while the WHOI engineers were working on battery replacement on the lander. The BPPT engineers also prepared the new design for a power supply system that utilizes higher DC voltage. They also sent one of their software engineers to update the application on the landing station computer, which forwards data from the sensor via satellite to the information read-down station in Jakarta.

Unfortunately, the result that the team most feared occurred: measurements exhibit a result that suggests that a short-circuit event in the submarine cable was possibly present, particularly near the shoreline, implying that a break on the remaining cable power in the submarine cable had occurred. The team could not precisely pinpoint the location of the break, however, because of the lack of a special tool for that purpose. Again, the team had no choice but to leave the site without having the whole system working completely.

Recommendations for Future Development

We learned valuable lessons through the four deployments and inspection missions to Siberut. All components of the prototype system were deployed, tested, and proved functional in an actual ocean environment, but they did not work together fully at the same time.

First, we learned that we should use highly professional and experienced technicians in submarine cable deployment. The cost to hire such professionals may be high, but this investment provides the assurance of successful cable deployment.

Second, in addition to hiring professional cable-laying engineers to ensure successful cable deployment, the team confirms that the use of a DP-equipped vessel is essential. In this project, hiring a DP-equipped vessel incurred extra financing beyond that allocated by the project funds, so the team attempted to use a non-DP-equipped ship at a cheaper cost. The outcome was not satisfactory. For the third deployment, the use of a DP-equipped vessel was essential. Fortunately, the team secured external funding to cover that cost. Even better and more encouraging for the team, the ship was equipped with a DP2 system, which is more advanced than a single DP system.

Third, the team learned the merits of a thorough process of canister fabrication, especially for canister design and testing. Those two aspects were not given sufficiently serious attention during the first and second deployments, which resulted in the failure of the second deployment. After two deployments that were designed to minimize costs, the canister was fabricated, tested, and deployed successfully in the third deployment.

Fourth, the use of proper protection armor for the submarine cable is essential. The submarine cable used in this project was a lightweight type that did not have sufficient protection against the harsh undersea environment and terrain. Cable with heavier armor costs more and was beyond the allocated budget, but if the team can obtain more external funding, purchasing a submarine cable with armor, at least with a single

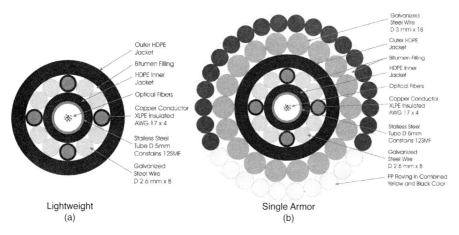

Outer HDPE Jacket
Bitumen Filling
HDPE Inner Jacket
Optical Fibers
Copper Conductor XLPE Insulated AWG 17 x 4
Stailess Steel Tube D 5mm Constains 12SMF
Galvanized Steel Wire D 2.6 mm x 8

Galvanized Steel Wire D 3 mm x 18
Outer HDPE Jacket
Bitumen Filling
HDPE Inner Jacket
Optical Fibers
Copper Conductor XLPE Insulated AWG 17 x 4
Stailess Steel Tube D 6mm Constains 12SMF
Galvanized Steel Wire D 2.6 mm x 8
PP Roving in Combined Yellow and Black Color

Lightweight
(a)

Single Armor
(b)

FIGURE 10-11. (a) Cross-section construction of lightweight cable. (b) Armored cable. (Diagrams by Xerandy and Iyan Turyana.)

layer of armor, is desirable. Figure 10-11 shows the construction of lightweight and armored type of submarine fiber-optic cable.

Finally, more detailed mapping ocean of the bottom is necessary. Detailed mapping would greatly assist the process of developing the cable-laying route and provide more detail of the terrain through which the cable would operate. Also needed to be considered in future budgets is the cost of operating and maintaining the landing station.

Chapter 11

Indian Ocean Tsunami Warning and Mitigation System: Initiation, Evolution, and Implementation

Harkunti P. Rahayu

The devastating impact of the 2004 Indian Ocean tsunami, with 230,000 people killed from Indian Ocean countries, was mainly due to the lack of a tsunami warning system in the region (Rahayu 2012). By 2004, the landscape of the region was characterized by several factors, including a lack of understanding of tsunami risk, very limited seismic observations to detect undersea earthquakes, very limited real-time sea level observations to verify incoming tsunami waves, no tsunami warning system to forecast tsunami threats, no national tsunami warning contact points to inform people in threatened communities, communities that were unaware and not prepared for tsunami risk, extremely limited response planning, and no coordinated international effort. As reported in table 11-1, by 2005 the status of the region's capacity to respond to a tsunami was very limited. The existing warning mechanism in the region was mainly for receiving teletsunami/distant tsunami bulletins transmitted from the Pacific Tsunami Warning Center and the Japan Meteorological Agency, with low capacity for a national warning center (31%), low capacity for receiving real-time seismic data (41%), and low availability of sea level data in real time or near real time at the central monitoring site. A teletsunami/distant tsunami is a tsunami with travel time above 2 hours from its originated source. Meanwhile, response procedures toward tsunami hazards hardly existed in the region. The 2004 event became a wake-up call for the establishment of a tsunami early warning system in the region.

Reflecting on the above issues, the Intergovernmental Oceanographic Commission of UNESCO (IOC-UNESCO) received a mandate from the international community to coordinate the establishment of an

Table 11-1. Comparison of 2005 and 2018 status of IOTWMS

	IOTWMS Status 2005		IOTWMS Status 2018	
Policies, plans, and guidelines	Legal framework in place for disaster warning formulation, dissemination, and response	59%	National tsunami policy in place	90%
			Local tsunami policy in place	60%
	National platform or other mechanism in place for guiding disaster risk reduction in general	94%	National tsunami disaster risk reduction plan in place	75%
	National Tsunami Warning and Mitigation and Coordination Committee or some other coordination mechanism in place	59%	Local tsunami disaster risk reduction plan in place	55%
			Community tsunami disaster risk reduction in place	40%
			National tsunami guidelines established	70%
	Disaster coordination mechanisms at community level established	75%	Local tsunami guidelines established	60%
	Tsunami emergency plans, tsunami evacuation plans, and/or signage exist indicating routes to safety or higher ground	9%		
Risk assessment and reduction	Tsunami hazard evaluation conducted prior to December 26, 2004	44%	Tsunami hazard assessment conducted	100%
			Tsunami risk assessment conducted	75%
	Historical record of past earthquakes and tsunamis documented	37%	Numerical modeling conducted for hazard assessment (Probalistic Tsunami Hazard Assessment and/or Deterministic Tsunami Hazard Assessment)	35%
	Tsunami vulnerability assessment conducted	22%		
	Numerical modeling studies conducted to calculate inundation from tsunamis	22%	Bathymetry used for tsunami hazard assessment	85%
	Accurate bathymetry and topography data exist for the coastlines	25%	Topography used for hazard assessment	80%

	IOTWMS Status 2005		IOTWMS Status 2018	
Detection, warning, and dissemination	International tsunami warnings received for tele-tsunamis from Pacific Tsunami Warning Center and/or Japan Meteorological Agency	94%	National capability to assess and/or receive potential tsunami threat information and advice and/or warn coastal communities	100%
	Agency receiving warnings staffed 24/7	94%	Warning center staffed 24/7	90%
	National or regional tsunami warning center to monitor and warn of regionally or locally generated tsunami in operation	28%	Access to national or international seismic networks	90%
	Warning center staffed 24/7	31%	Access to national or international sea level networks	85%
	Real-time seismic data received	41%		
	Sea level data available in real time to the central monitoring site or available in near real time	41%		
Standard operating procedures	Local government disaster preparedness and emergency response assessed	59%	Warning dissemination standard operating procedures (SOPs) in place	90%
	Community and ordinary citizen disaster preparedness and emergency response assessed	25%	Evacuation call SOPs in place	80%
	Response procedures for regional or locally generated tsunami in place	19%	Community evacuation SOPs in place	60%
Tsunami exercises	Response procedures have been tested or exercised	19%	Tsunami exercises conducted at national level	70%
	Public is aware of what a tsunami is and how to respond to both locally generated and distant tsunamis	37%	Tsunami exercises conducted at regional level	55%
			Tsunami exercises conducted at city level	35%
			Tsunami exercises conducted at village level	50%
			Tsunami exercises conducted at community level	50%
			Tsunami exercises conducted at school level	30%
	IOTWMS Status 2005		IOTWMS Status 2018	

Continued

TABLE 11-1 *continued*

			Tsunami-related education and awareness material	
Awareness, preparedness, and response	Community-level education and preparedness programs for national hazards or tsunami exist	47%	- Leaflets or flyers	65%
			- Posters	70%
	Tsunami education and public outreach program in place	6%	- Booklets	60%
	Earthquake and tsunami hazards and preparedness is incorporated into educational curricula for schoolchildren	12%	- Information boards	30%
			- Tsunami signage	25%
			- Video or other visual/oral media	65%
			- Indigenous knowledge	35%
			- Teaching kits	50%
			- School curricula	45%
			- Public evacuation maps	25%
Media training	Training programs for the media on tsunami hazards, mitigation, warning, and preparedness exist	22%	Media arrangement SOPs in place	80%

Source: UNESCO/IOC 2020.

entirely new warning and mitigation system for the Indian Ocean region during the course of several international and regional meetings, including the Special ASEAN Leaders' meeting (adopted in Jakarta, January 6, 2005); UN Conference on Small Island Developing States (Port Louise, Mauritius, January 10–14, 2005); Common Statement of the Special Session on Indian Ocean Disaster and the Hyogo Framework of Action 2005–2030 adopted at the UN World Conference on Disaster Reduction (Kobe, Japan, January 18–22, 2005); the UN General Assembly Resolution 59/279 (adopted in New York, January 19, 2005); the Phuket Ministerial Meeting on Regional Cooperation on Tsunami Early Warning Arrangements (Phuket, Thailand, January 28–29, 2005); the Group on Earth Observations (GEO) Communique (adopted in Brussels, January 16, 2005); and several technical meetings held in India, China, and Indonesia. Then, during the twenty-third Session IOC Assembly (Paris, June 21–30, 2005), the Indian Ocean Tsunami Warning and Mitigation System was formally established through Resolution IOC-XXIII-12 (Intergovernmental Oceanographic Commission 2005). The system is known as UNESCO Intergovernmental Oceanographic Commission/ Intergovernmental Coordination Group Indian Ocean Tsunami Warning and Mitigation System (UNESCO IOC/ICG IOTWMS), called IOTWMS, with twenty-four active member states from a total of twenty-eight member states. It was formed as a coordinated network of national systems and capacities and is part of a global network of early warning for all ocean-related hazards: ICG/Caribe-EW, with thirty-two member states and sixteen territories; ICG/NEAMTWS–Northeastern Atlantic, the Mediterranean, and connected seas, with thirty-nine member states; and ICG/PTWS Pacific Tsunami Warning and Mitigation System, first convened in 1968, with forty-six member states.

The two basic concepts of early warning systems, end-to-end and people-centered, were adopted in the development of IOTWMS to build a sound and solid system, covering hazard detection and forecast, threat evaluation and alert formulation, alert dissemination and public safety messages, and preparedness and response; see figure 11-1. The four elements of a people-centered early warning system, endorsed by the United Nations International Strategy for Disaster Reduction Third Early Warning Conference in Bonn 2006 and described in chapter 2, consist of risk knowledge, monitoring and warning service, dissemination and communication, and response capability. These four elements are embedded in the end-to-end system, shown in figure 11-1.

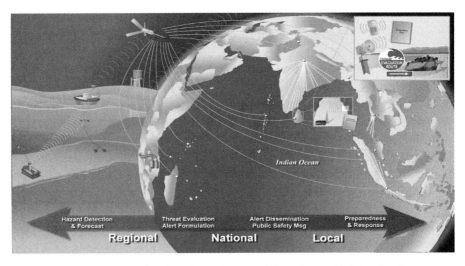

FIGURE II-I. End-to-end Indian Ocean Tsunami Warning and Mitigation System.

Evolution of the Indian Ocean Tsunami Warning and Mitigation System Since 2005

After the initial stage of development in 2005, the IOTWMS evolved into a "system of systems" concept, that is, a network with multiple tsunami service providers (TSPs) and twenty-four national tsunami warning centers (NTWCs). The TSPs are expected to generate real-time products for all NTWCs within the Indian Ocean region. With existing capacity and high commitment, India (with its National Center for Ocean Information Services), Indonesia (with its Meteorological, Climatological, and Geophysical Agency), and Australia (with its Bureau of Meteorology) have taken the lead to become TSPs for the region. During the process of developing the TSPs, the Pacific Tsunami Warning Center (PTWC) and the Japan Meteorological Agency (JMA) voluntarily provided tsunami warning services to all NTWCs of the Indian Ocean region. The IOTWMS considered the PTWC and the JMA as interim advisory service providers until the completion of the TSP function. After that, the NTWCs of the member states are solely responsible for providing warnings and alerts to their citizens based on their own analyses of the situation.

The tsunami models that the TSPs run to produce their libraries or database of tsunami scenario forecasts are similar with respect to governing equations and physics, but they differ in their configurations. Within the bounds of the common methodology described above, they differ in many technical details of their forecasting process, such as the location and spacing of earthquake scenarios on which the databases of model forecasts are based; the assumptions about earthquake rupture geometry and energy transfer to the ocean; the models, analysis, and overall processes; and the method of selecting representative values for each forecast zone. To develop a uniform warning mechanism through tsunami wave modeling and forecasting, however, the three TSPs had to establish a common template for the entire Indian Ocean coastline (Hettiarachchi 2018). As shown in figure 11-2, the information flow from seismic detection and sea monitoring, combined with model results, was processed by each TSP using its own mechanism to produce threat information and warning status to all twenty-four member states' NTWCs. The NTWCs are then responsible for disseminating the warning to their national disaster management offices and local disaster management offices that issue alerts using their own downstream warning chain infrastructures.

As also shown in figure 11-2, significant progress has been achieved since 2005. In 2011, the three TSPs (Australia, India, and Indonesia)

Modus Operandum
Tsunami Detection, Warning and Dissemination
IOTWMS

unesco
Intergovernmental
Oceanographic
Commission

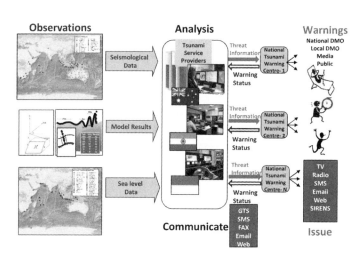

FIGURE 11-2. Indian Ocean Tsunami Warning and Mitigation System Modus Operandum: tsunami detection, warning, and dissemination.

moved into parallel operation with the interim advisory services (the Pacific Tsunami Warning Center and the Japan Meteorological Agency). For the first time, a standard operating procedures (SOP) training workshop, an Indian Ocean–wide wave exercise (IOWave11), and a communication test were conducted to assess the performance of the end-to-end system. In addition to strengthening the capacity of IOTWMS member states, these activities are now conducted on a regular basis. For example, the IOWave exercise is conducted in every two years, the SOP training workshop is conducted annually, and communication tests are performed every six months. The aim of the SOP training workshops is to enable the twenty-four NTWCs to develop their own national SOPs based on the those of the three TSPs. Meanwhile, the communication devices for warning dissemination are tested to validate the TSPs' performance for dissemination, the NTWCs' reception of the warnings, and their feedback to the TSPs. The IOWave exercises are intended to test the readiness of the end-to-end system, as well as the resilience and responsiveness of the community at risk. Since 2005, six IOWave exercises have been conducted: in 2009, 2011, 2014, 2016, 2018, and 2020. The IOWave20 exercise was conducted during the COVID-19 pandemic and thus it tested only the readiness of the communication system, with no movement of people for evacuation.

At beginning of the IOTWMS's development, very few real-time seismometers and almost no real-time earthquake monitoring networks were operating in the Indian Ocean. After more than fifteen years of development, however, several countries now operate real-time seismic networks; the three TSPs and a few NTWCs are also capable of estimating earthquake parameters in near real time (less than 10 minutes after the event). Prior to 2004, there were only four real-time tide gauges in the Indian Ocean. Today, several countries operate real-time bottom pressure recorders and tide gauge networks, and the three TSPs and a few NTWCs are capable of monitoring real-time sea level data (tsunami confirmation within 30 to 60 minutes) (ICG/IOTWMS 2019).

The Governance of the Indian Ocean Tsunami Warning and Mitigation System

During the establishment of the IOTWMS in 2005, the IOTWMS was supported by six working groups, covering the three principal pillars of a warning system: risk assessment and reduction (collect data and undertake tsunami hazard and risk assessment); detection, warning, and dissemination (develop hazard detection, monitoring and early warning services, and communicate threat information and early warnings); and awareness and response (build national and community response capabilities) with the support of the secretariat based in Perth, Australia, to coordinate all activities of the ICG. Later in 2012, these six working groups were merged into three, representing the pillars of a warning system, as stated above. For effectiveness in governance, however, these three working groups were merged into two—Working Group 1 on Tsunami Risk, Community Awareness and Preparedness and Working Group 2 on Tsunami Warning, Detection and Dissemination—at the 2015 ICG meeting in Oman. Thus, Working Group 1 covers activities related to increasing capacity for hazard assessment; risk assessment; use of hazard and risk assessment on disaster risk reduction policies, plans, and guidelines; and emergency response policies, evacuation planning, tsunami exercises, education, and community awareness. Meanwhile, Working Group 2 has responsibility for increasing capacity for detecting, monitoring, and analysis of tsunami threats; organizational performance to function 24/7; tsunami early warning system thresholds; and tsunami early warning system downstream warning chains.

The governance of ICG/IOTWMS is shown in figure 11-3. The Working Group will work closely with the Task Team established for certain purposes for certain periods, such as the task teams for IOWave, Capacity Assessment for Tsunami Preparedness, Tsunami Preparedness for Near-Field Tsunami Hazards, and Scientific Hazard Assessment of the Makran Subduction Zone. The governance of IOTWMS is dynamic and compliant with current and future needs. The endorsement is taken during ICG meetings, which are held in every two years. For example, to achieve the target of the Ocean Decade Tsunami Program, a new working group and several task teams— the Tsunami Ready Working Group, the Critical Infrastructure Task Team, and the Atypical Tsunami Task Team—are needed.

Gaps and Achievements of IOTWMS

Since 2005, only two capacity assessments have been conducted. The first was done in 2005 to measure the baseline data on capacity of the three pillars to be used for the development plan of IOTWMS. Then, in 2018, after thirteen years, the second capacity assessment for tsunami

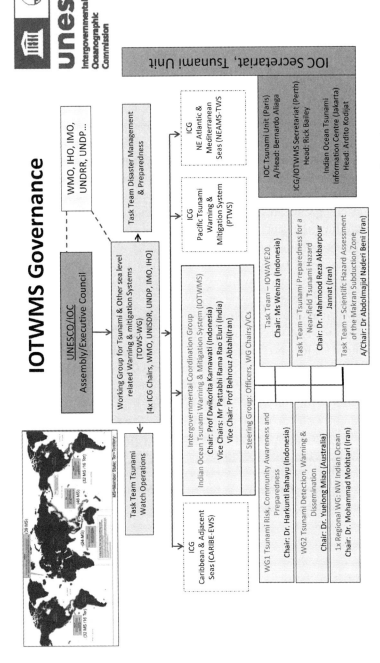

FIGURE 11-3. Indian Ocean Tsunami Warning and Mitigation System Governance (Karnawati 2021).

preparedness was conducted through an online survey, aiming to provide a new baseline for tsunami preparedness capacity in the region for further improvement. The survey included five sections: (1) basic information; (2) policies, plans, and guidelines; (3) risk assessment and reduction; (4) detection, warning, and dissemination; and (5) public awareness, preparedness, and response (Rahayu 2021).

Of the twenty-four active IOTWMS member states, a greater number of them (twenty-one) responded to the 2018 assessment than responded to the 2005 assessment (sixteen), and in-depth assessment results indicated that there has been considerable improvement across all components of the IOTWMS since 2005. The 2018 assessment also identified specific gaps and priorities for setting capacity development requirements at both regional and national levels (UNESCO/IOC 2020). Table 11-1 summarizes the 2005 and 2018 capacity assessment surveys, comparing the status of IOTWMS in 2005 to that in 2018. The percentage columns refer to the percentage of countries participating in each survey answering "yes" to the related question, with a "partial yes" in the 2005 assessment counted as a "half yes." The 2005 percentages are based on responses from sixteen countries, and the 2018 percentages are based on responses from twenty-one countries, with fourteen countries in common. Given the differences between the two assessments, the table is intended to provide a broad comparison only to indicate the scale of capacity improvement in the IOTWMS since 2005.

To Address Capacity Gaps and Support Requirements

Based on table 11-1, a summary of the capacity gaps and support requirements that emerged from the 2018 Indian Ocean capacity assessment of tsunami preparedness is presented in table 11-2. This summary is intended to provide recommendations for future capacity development activities in the Indian Ocean region to achieve the objectives of the Medium Strategy of IOTWMS 2019–2024 (UNESCO/IOC 2019), namely:

1. An interoperable tsunami warning and mitigation system based on coordinated member state contributions that uses best practices and operational technologies to provide timely and effective advice to NTWCs.
2. Indian Ocean communities at risk are aware of the tsunami threat; are able to assess and reduce risk; and are prepared and ready to act to save lives, protect property and infrastructure.

TABLE 11-2. ICG/IOTWMS strategic pathway in context of the UN
Ocean Decade

IOTWMS Pillar	Ocean Decade Tsunami Program
1. Risk assessment and reduction: *Hazard and risk identification and risk reduction*	• Access to data, tools, and communication platforms and to protocols and training to timely and effectively warn coastal and maritime communities
2. Detection, warning, and dissemination: *Rapid detection and warning dissemination down to the last mile*	• Expansion of existing and deployment of new technologies addressing observational gaps • Wide expansion of real-time and near-real-time data access and availability • Access to data, tools, and communication platforms and to protocols and training to timely and effectively warn coastal and maritime communities
3. Awareness and response: *Public education, emergency planning, and response*	• 100% of communities at risk of tsunami prepared for and resilient to tsunamis by 2030 through the implementation of the UNESCO/IOC Tsunami Ready Program

There has been major improvement in pillar 1, risk assessment and reduction. Fewer than half of the countries assessed in 2005 had conducted tsunami hazard assessments, and fewer than one-fourth of the countries had done numerical modeling of tsunami inundation. In contrast, by 2018 all the countries assessed had conducted tsunami hazard assessments, with the majority as part of multihazard assessments, as also shown in figure 11-4 (UNESCO/IOC 2020), which is available in the online directory for this book, https://islandpress.org/tsunami.

There is still strong need to strengthen pillar 1 by several recommended actions to increase (1) engagement of other national, regional, or international actors in carrying out tsunami hazard and risk assessments; (2) availability of publicly accessible data for tsunami hazard and risk assessments; (3) capacity for tsunami hazard assessment, especially in the areas of evacuation mapping, hazard mapping, and inundation mapping; (4) capitalization of the existing capacity of member states to deliver training on hazard mapping and inundation mapping; (5) capacity for city-, village-, and community-level tsunami risk assessments; and (6) capacity for developing products from tsunami risk assessments, such as risk maps, evacuation maps, guidelines, and action plans.

For pillar 2, *detection, warning, and dissemination,* the 2005 baseline assessment showed that nearly all the assessed countries had an agency for receiving international tsunami warnings from PTWC or JPA and were staffed 24/7. A few countries had a national agency for monitoring and warning their citizens of regionally or locally generated tsunamis. In contrast, the 2018 capacity assessment survey shows that all countries reported the capability to assess or receive potential tsunami threat information from IOTWMS TSPs or to use a combination of TSP data and their own threat assessment data to provide advisories or warnings to their coastal communities. There is still a strong need, however, to increase the capacity for (1) analyzing real-time seismic and sea level data for tsunami threat, (2) tsunami modeling to support generation of threat forecasts, and (3) further study to examine whether there is a need for so many different software tools to be used to analyze data for tsunami threat or tsunami modeling. Increasing the frequency of tabletop or similar tsunami warning exercises to review and test SOPs is essential, as is reducing the complacency among countries that have not experienced a recent tsunami event.

For pillar 3, *awareness, preparedness, and response,* the 2005 assessment showed that the existence of SOPs was not explicitly addressed; only related awareness and response procedures were assessed. In comparison, the 2018 capacity assessment indicates that eighteen countries have developed SOPs for their upstream operations. Most of these countries have also developed SOPs for their downstream operation, such as for warning dissemination, communications with the NTWC and other stakeholders, evacuation call procedures, and media arrangements. Moreover, thirteen countries have developed SOPs for community-level evacuation. Several actions, however, are still required to improve the SOPs at the interface between upstream and downstream notification of warnings, including the operation of a 24/7 emergency operations center, receiving information from the NTWC, providing response criteria and decision making, and allocating the associated human resources and infrastructure to these tasks. Further, supporting actions are essential to provide support for the development of community-level evacuation SOPs; capitalize on the willingness of countries to share their SOPs and good practices across member states; provide training and share member states' experience regarding different types of evacuation infrastructure; incorporate tsunami exercises into cities, villages, communities, and schools; provide training and share member states' experience regarding different public engagement materials; develop educational materials such as teaching kits and encourage the incorporation of tsunami awareness into the school curricula; and raise awareness of the Global Disaster Risk Reduction Day (October 13) and World Tsunami Awareness Day (November 5).

To complement the three pillars of IOTWMS, the 2018 capacity assessment has identified the component of *policies, plans, and guidelines.* Findings from the assessment show that there are strong needs for support to increase the availability of tsunami policies, plans, and guidelines in the prevention and mitigation, preparedness, and recovery and reconstruction phases of disaster management. Also needed is support to increase availability of tsunami policies, plans, and guidelines at the local level, either as stand-alones or as part of a multihazard approach.

Conclusion

As envisioned by the UN Decade of Ocean Science for Sustainable Development 2020–2030, a safe ocean is where people can continue to interact with the sea and oceans for their livelihood and lifestyle without fear of coastal hazards (e.g., tsunamis), as the hazards are well understood and observed, where the potential impact is predicted timely and accurately, and where the people are already equipped with awareness, knowledge, preparedness, skills, and abilities to safeguard themselves in the case of an emergency (Karnawati 2021). Given the tsunami program's framework for action to protect communities from the world's most dangerous wave, the IOTWMS provides a strategic pathway to achieve the vision of a safe ocean by implementing the three pillars of its Tsunami Ready program.

Pillar 1, hazard and risk identification and risk reduction, will focus on increasing access to data, tools and communication platforms, protocols, and training to timely and effectively warn coastal and maritime communities. Pillar 2, rapid detection and warning dissemination down to the last mile, will focus on three activities: expansion of existing and deployment of new technologies addressing observational gaps; wide expansion of real- and near-real-time data access and availability; and access to data, tools and communication platforms, protocols, and training to timely and effectively warn coastal and maritime communities. Pillar 3, public education, emergency planning, and response, will target a goal of enabling 100 percent of communities at risk of tsunami to be prepared for, and resilient to, tsunamis by 2030 through the implementation of the UNESCO/IOC Tsunami Ready program or other similar programs using participatory engagement. Such programs would include engagement from national and local government agencies, research and development agencies, universities, other scientific organizations, commercial and business entities, nongovernment platforms and organizations, and community.

References

Hettiarachchi, Samantha. 2018. "Establishing the Indian Ocean tsunami warning and mitigation system for human and environmental security." *Procedia Engineering* 212: 1339–46.

Intergovernmental Oceanographic Commission. 2005. *Intergovernmental Coordination Group for the Indian Ocean Tsunami Warning and Mitigation System (ICG/IOTWS).* Intergovernmental Oceanographic Commission (IOC) of UNESCO. Resolution XXIII-12. UNESCO. Twenty-Third Assembly, Paris, June 21–30, 2005, IOC-XXIII/3, Annex II, 18.

Karnawati, Dwikorita. 2021. "Chair's report to Intersessional Meeting of the Intergovernmental Coordination Group for Indian Ocean Tsunami Warning and Mitigation System (ICG/IOTWMS)." November 23–24, 2021.

Rahayu, Harkunti P. 2012. "Integrated logic model of effective tsunami early warning system." Kochi University of Technology, Japan.

Rahayu, H. P. 2021. "Tsunami risk, awareness, and preparedness report, Chair, Working Group 1." *Report to Intersessional Meeting of ICG/IOTWMS*, Paris, November 23–24, 2021.

UNESCO/IOC. 2019. *Indian Ocean Tsunami Warning and Mitigation System (IOTWMS): Medium Term Strategy, 2019–2024.* Paris: UNESCO. Technical Series No. 144.

UNESCO/IOC. 2020. *Capacity Assessment of Tsunami Preparedness in the Indian Ocean–Status Report, 2018.* Paris: UNESCO, IOC Technical Series No. 143.

Chapter 12

Creating a Sustainable Learning System in Regions of Risk

Louise K. Comfort, Wahyu W. Pandoe, Harkunti P. Rahayu, and Iyan Turyana

Returning to the question of how communities learn to recognize risk and act collectively to reduce loss from extreme events, we review insights gained from the design, implementation, and testing activities undertaken and documented in the preceding chapters. The goal remains the same—to mobilize an entire community quickly to reduce harm from extreme events—but the requirements for doing so extend from building and implementing sociotechnical strategies to reduce known risk to the larger, more difficult task of sustaining an adaptive learning system in a community exposed not only to known risk, but also to risk that is yet unknown.

We recognize that the demands of protecting communities from urgent, large-scale, compound risks require a range of technologies and social innovations to build collective cognition and action, yet there are limited indicators of when or where such hazards may occur and with what degree of severity. The devices designed, tested, and validated in the preceding chapters offer a sociotechnical format to bridge communications among different groups, organizations, and jurisdictions to mobilize actions across distance and time. These devices rely on the human capacity to learn and to advance knowledge through the search for improved methods of managing uncertain events and updating methods and instruments based on learned experience in actual environments.

In the previous chapters, we presented a series of learning experiences among multiple organizations and jurisdictions for communities exposed to risk, supported by technologies that could assist them in reducing uncertainties that characterize one type of hazard, tsunamis. Chapter 1 introduced the design of a bowtie architecture for a sociotechnical information system to facilitate the collection, aggregation, analysis, and exchange of information among multiple organizations and jurisdictions in near-real time as an urgent event emerges and develops. Chapter 2 provided the organizational design for, and an assessment of, the Indonesian Tsunami Early Warning System (InaTEWS) currently in practice, noting strengths and weaknesses of the current system. Chapter 3 built on that assessment of InaTEWS to introduce the technical design for a bowtie information system that would address a gap in specifically the "last mile" communication of warning of an imminent tsunami to neighborhood residents. Chapter 4 extended the design of a neighborhood network to introduce low-cost connections to the Iridium satellite system that enables wireless communication among local operations personnel and neighbors organized through a community network, essentially creating a movable wireless "hot spot" that is led by trusted local leaders and facilitates communication among neighborhood residents as they move to safe ground. Chapter 5 presented the engineering design for extending small neighborhood *mushollahs*, or small mosques, to become shelters in case of tsunami inundation. The extensions would be built over time according to professional engineering standards, with neighbors contributing both skills and money to provide safe shelter in their immediate locales. That chapter highlights the contribution of local professional engineers guided by Andalas University to enhance the capacity of neighborhoods to create their own safe spaces from potential tsunami inundation.

Importantly, the technical designs presented in chapters 3, 4, and 5 are simulated and tested to validate their performance before implementation. Chapter 6 presented a simulation of the development of "situational awareness" or cognition of risk information when it is transmitted through a bowtie-designed information system. The findings show a rapid increase but then a surprising drop before community residents proceed, as recipients stop to check the information they received and then move to action. Chapter 7 presented a systematic test of the entire neighborhood network presented in chapters 3, 4, and 6. The results validate the basic assumptions underlying the design for a neighborhood network as a sociotechnical system to support local evacuation or sheltering at the neighborhood level in event of a sudden, urgent tsunami.

Chapters 8, 9, and 10 presented the design, implementation strategies, and results for the undersea network, with models, engineering specifications, and an account of obstacles that emerged and were overcome as the network was deployed in the Mentawai Sea near Siberut Island. These chapters presented invaluable insights gained from this experience for strengthening the design and future implementation of a fully developed undersea acoustic network that could be integrated into the national InaTEWS. The two networks—neighborhood and undersea—would strengthen the valid confirmation of early tsunami detection and delivery of tsunami warning to the local neighborhood residents, the initiating and ending points in the InaTEWS that are most vulnerable to error.

The rationale for this book is to extend the insights gained from the learning experiences presented in these chapters to the wider Indonesian society, as well as to other nations exposed to tsunami hazards, to inform actions toward more comprehensive mitigation and community resilience. The chapters document a sociotechnical framework and set of instruments to enable individuals, organizations, and jurisdictions to adapt and adjust their recognition of, and response to, risk as a dynamic process that changes over time.

Integration of Qualitative and Quantitative Measures of Risk

Fundamental to building community resilience to risk is determining who is at risk from what hazard and how this risk could be averted or lessened. This task involves measurement of risk as an undefined, moving target that defies standard linear measurement. The basic conditions in which risk occurs become the background that can be identified and measured, so the process of measurement focuses on the rate of change in conditions that lead to risk. Necessarily, that means employing a mix of quantitative measures to document existing conditions, but projecting qualitative assessments of conditions, actors, and processes that are dynamic. The focus is on a clear articulation of the goal of a resilient community and what is required to reach it, while the methods of doing so may vary among individuals, groups, institutions, and jurisdictions that make up that community and support its ongoing efforts to reduce risk.

Although measuring the status of known conditions—depth of seafloor, population of coastal communities, historic record of seismic movement in the study region, distance to shelters and safe ground in the selected community—provides a status report on the degree of known risk, monitoring those conditions systematically provides a means of estimating the rate of change that may indicate a sudden, urgent event. Such measurement cannot eliminate uncertainty in a dynamic world, but it can indicate the probability for an extreme event to occur and, importantly, underscore the merits of considering alternatives for action should such an event occur. Alerts generated by systematic monitoring of changing conditions that lead to risk can be powerful indicators for action in a complex, dynamic world. Building a systematic knowledge base of conditions prone to risk prior to a hazard event to facilitate rapid assessment of probable risk is a step toward community resilience.

In complex adaptive systems, measurement necessarily takes a different form from analytical processes in standard hierarchical organizations. Multiple components in complex systems interact and influence one another to generate new and unexpected dynamics. Understanding change as a process that allows interacting components to evolve in novel ways to fit their operating environment more appropriately shifts the process of measurement to identifying the networks of actors and the technical instruments they use as sociotechnical mechanisms engaged in managing risk. Devising sociotechnical networks in novel environments, such as the neighborhoods of Padang City or the undersea acoustic network in the Mentawai Sea, to identify interactions among human actors, technical instruments, and changing environmental conditions represents an innovative contribution to enabling communities to recognize and act on perceived risk in timely, valid ways. Consequently, measurement of the interacting variables becomes nonlinear in the effort to capture the dynamic exchanges between community agents, the technical infrastructures available to them, and their local and regional physical contexts. Devising measurements, such as shown in chapters 3, 4, and 6, and defining data collection methods that assess the rate of change at different scales and times of operation to capture the dynamics of change in the whole system are central to managing risk in real time.

Integrating data from multiple methods of assessment and aligning the results to provide a coherent profile of risk to inform action for the whole community becomes a multilevel task. This task includes creating communication linkages and professional interaction within and between operational levels of organizations, disciplines, and jurisdictions that are involved in early tsunami detection and warning. The primary goal is to enable a system-wide awareness of tsunami risk and identify the steps that can be taken by different groups at different locations to mobilize timely, informed action to move to safe shelter.

The Learning Process in Complex Adaptive Systems

Learning occurs in complex adaptive systems, but at different rates among different groups in different locations. Designing a learning process to facilitate learning among multiple groups in a single community that have different backgrounds, experience of risk, knowledge of resources, and commitment to manage risk for the whole community is a complex task. Specifically, designing a learning process to complement the construction and testing of novel sociotechnical networks for early tsunami detection and warning means incorporating stages for demonstration, explanation, review, and revision among groups with different disciplinary backgrounds, levels of responsibility to the community, experience with hazard risk, and commitment to protect the whole community from harm. It means viewing a community that is exposed to risk as a complex adaptive system with the capacity to operate as a coherent entity rather than as a collection of disconnected individual actors or groups moving at different rates in multiple, often conflicting directions. It assumes the interest, commitment, and capacity of a leadership group within the community to develop collective intelligence that will inform coordinated action.

At least four distinct challenges are involved in achieving a viable learning process that is sustainable over time. The first is to articulate a goal for the whole community that is clearly understood and supported by the multiple organizations and actors who make up the community as a complex adaptive system, which means recognizing that there will be differences in perception of risk, capacity to act, and resources to allocate for the implementation of that goal among these different components. Still, the goal, clearly stated and understood, creates a unifying vision for community members to accept as their primary rationale for responsible action. The goal to improve early tsunami detection and warning to save lives in coastal cities has animated this research, first articulated by an international interdisciplinary group of researchers and practicing managers in 2005 after the trauma of the 2004 Indian Ocean tsunami and redefined by a broader group of investigators in 2013.[1] This goal has sustained the project throughout the years of study, design, trial, evaluation, redesign, and continued effort to produce a tested prototype for actual implementation in Padang City and the Mentawai Sea.

Importantly, the goal of risk reduction is not imposed on community members through a hierarchical structure, but rather offered as an opportunity for shared development. Some members may respond more directly than others, but the power of a shared goal is the willing commitment of members of the system to acknowledge their responsibility to one another as well as to the "system" they are shaping by their collective actions.

The second challenge is to identify the specific individuals, groups, organizations, agencies, and jurisdictions that constitute the principal actors of the system. This task sets the learning process directly in the context of an actual community with its leaders, followers, opponents, and arbiters regarding shared risk and its specific resources that are affected by the hazard. It also illustrates the interconnections, both positive and negative, among the actors in the system and the need to identify the feedback loops within and among the social and technical components of the system. These feedback loops generate different responses among different actors, some positive, some negative. Advanced technologies are central in both identifying the feedback loops and noting their strength in enhancing or limiting coordination among different groups in the community for early tsunami preparedness and warning.

For example, the capacity of the neighborhood volunteer groups, or KSB, in Padang, identified in chapter 2, could be substantively enhanced as primary actors in local preparedness and evacuation strategies by the local electronic network presented in chapters 3, 4, and 6. In addition, the timeliness, validity, and accuracy of the early tsunami detection and the warning information transmitted to the KSB groups would be validated by data transmitted from the undersea network to BMKG (Indonesian Agency for Meteorology, Climatology, and Geophysics) and further to the provincial and municipal emergency preparedness offices, as detailed in chapters 8, 9, and 10. These key information linkages activate social groups through accessible technical infrastructure, advanced knowledge and skills, and social empathy.

The third challenge is to model strategies for mobilizing action among the different actors, groups, and types of technical infrastructure that balance potential conflicts and lead broadly toward the goal of safe protection from harm articulated for the whole system. Multiple strategies may be necessary to adapt to different capacities among different components at different stages in the learning process. School children, for example, will need a different level of guidance and care than adults; businesses engaged in the manufacture and distribution of products will require levels of protection for both goods and customers; rural residents may have less access to communications technology and be more skeptical of recommendations for evacuation from distant entities with whom they are not familiar. To meet this challenge, the exercise of simulating possible strategies as exercises in evacuation in event of tsunami threat can be instructive, as was reported in chapters 4, 6, and 7.

The fourth, and critical, challenge is the alignment and integration of the strategies outlined in the previous three challenges as essential to enable a broad learning process to fulfill the stated system-wide goal of resilience to tsunami risk. This step is the most difficult as it requires, first, aligning the expertise to design and develop both technical and organizational infrastructures with the legal authorities to implement the strategies in actual locations with the financial and practical resources necessary to do so. Further, it requires integrating the knowledge gained from each strategy implemented into a coherent account that enables multiple audiences—policy makers, public officials, business owners, neighborhood leaders, and others—to understand the changes and make their own commitments to support a system-wide effort for, in this study, early tsunami detection and warning. This fourth challenge represents the culmination of the earlier steps and leads to a sustainable process of learning across individuals, groups, organizations, infrastructures, and jurisdictions over time.

Key Insights Gained to Support Early Tsunami Detection and Warning Systems

Reviewing the findings from the previous chapters, we summarize key insights gained from the documented learning experiences that are relevant for the adoption of the prototype early tsunami detection and warning system into the official InaTEWS or other tsunami mitigation and warning systems. Based on the experience of our international, interdisciplinary research and implementation team, we find the following set of interrelated insights constructive for the extension of the Hazard SEES prototype into full implementation and operation in other locations prone to tsunami risk.

Specification of goal, operational requirements, and costs

Although careful specification of operational requirements and costs needed to achieve the project's goal is relevant to the conduct of any complex, interdisciplinary project, it is fundamental to projects that are undertaken in novel working conditions. A major subgoal of this project was to find economical means of designing and implementing sociotechnical infrastructures for the project. Yet, especially in instances where the actual conditions were not well known, efforts to save funds resulted in the need to redesign and reinvest in more robust equipment and more experienced operators. For example, the decision to use a small, local ship with a crew inexperienced in working in deep water to lay the undersea cable for the first deployment of the undersea acoustic network in early January 2020 resulted in failure of the cable to reach shore and the need to redo the deployment. The initial decision was made to stay within the allocated budget, but in practice, it doubled the cost by requiring the purchase of a new undersea cable and hiring a professional engineering company with the proper equipment to operate in deep ocean waters. More expensive, yes, but also more effective and more efficient in anticipated time.

Managing external conditions of financial resources, organizational change, and professional support

Complex, novel operations that span geographic distance, multiple organizations, jurisdictions, and time require building the internal capacity of an interdisciplinary, international team to conduct the operations. In this operation, the legal requirements for conducting a major international experiment in both a local city and in nationally controlled waters meant securing the authority, approval, and financial support from several major agencies of the Indonesian government. Although approval and commitment to financial support had been secured from relevant Indonesian government agencies prior to the beginning of the project, the democratically elected government changed in 2014 to place new officials in authority with new priorities for managing risk.

The necessary support from Indonesian agencies to support the project was secured, but it took time and effort to explain to new officials, demonstrate the concept of operations, and document the benefits that would accrue not only to Indonesia but to other countries exposed to tsunami risk. Including provision for time and effort to maintain collaborative relationships among all entities involved in conducting a complex, interdisciplinary, international project is essential.

Managing uncertain, risky working conditions

A deep-water ocean environment poses risky working conditions and requires professional-grade vessels, equipment, and experienced technicians to execute a novel design. The execution of the prototype undersea acoustic network to operate in waters at a depth of 700 meters had never been done before, and consequently, investment in advanced vessels, equipment, and skilled technicians allows a range of adaptation to unanticipated conditions that enables successful management of unknown risk. For example, an experienced BPPT deputy chairman/ocean engineer specified that a ship with a dynamic positioning system was essential to enable the engineering team to do the delicate work of replacing the canister at sea. Such a ship was much more expensive than the budget for the project allowed. Given informed knowledge and recommendations, it was necessary to reallocate existing budget and time and find new financial resources to make this expenditure. The investment paid off when, in the December 2021 deployment, the crew discovered that the cable connected to the canister was twisted in the retrieval process, but using the ship with a dynamic positioning system, the crew was able to maneuver the ship to untwist the cable and successfully retrieve and replace the canister. This experience demonstrated that investment in professional-grade vessels, equipment, and technicians in the initial effort facilitates adaptation to unexpected conditions and leads to successful outcomes with less delay, less redesign, and less costly rework.

Testing in all phases of design, development, and implementation of novel systems

Again, this insight is relevant to any major complex project, but it is especially relevant to projects undertaken in novel conditions when the interactions between newly designed instruments and the actual conditions in which they will be operating have not been tried before. This insight was documented by the experience of designing the canister for the receiver modem for the undersea acoustic network. In July 2020, the successful installation of the new undersea cable was marred by an apparent water leak in the canister that had not been fully tested. In the December 2021 deployment, the canister went through repeated tests and returned successful results each time. Yet, in the April 2022 deployment, the undersea cable that had tested successfully in December 2021 had apparently developed a break in the circuit. This result revealed that the light cable chosen to minimize costs for the prototype development could not withstand the continued pressure of a deep-ocean environment. This experience demonstrated that investment in an armored cable, even for a short, 7 kilometer distance, is essential to maintain operations of the acoustic network anchored in rugged terrain in deep water and that testing, even after a successful operation, is essential for continued performance in novel environments.

International, interdisciplinary collaboration to address global problems

Tsunami risk is a major global problem, and this project documented the invaluable contribution of international, interdisciplinary collaboration in addressing this risk. Although Indonesia is exposed to major seismic risk that generates earthquakes, tsunamis, and volcanic eruptions, other nations are also vulnerable to the death and destruction that result from such extreme events. Under the United Nations Charter, all 192 member nations have an obligation to assist a fellow member nation that has suffered from a destructive event. In this instance, collaboration between Indonesia and the United States was initially supported by the US National Science Foundation but extended by the Swiss Re Foundation from Zurich, Switzerland, and further advanced by the Yayasan Anak Bangsa Bisa (YABB) Foundation in Indonesia. These international institutions provided external funding for this global problem, while BPPT, the Indonesian governmental agency responsible for the design and maintenance of a buoy system for early tsunami detection and warning, provided continuing expertise, personnel time, vessels and crew, and knowledge of local conditions to the implementation tasks. Importantly, the interdisciplinary requirements for expert knowledge to design and carry out the project drew researchers from public policy, urban planning, computer science, ocean engineering, seismology, and structural engineering from four US research institutions, University of Pittsburgh, Carnegie Mellon University, Northwestern University, and Woods Hole Oceanographic Institution, as well as from the research centers and disciplinary departments of Indonesian universities ITB and Andalas University. These institutions provided continuing support through faculty expertise, student participation, and local knowledge throughout the years of the project.

This prototype early tsunami detection and warning project documents in multiple ways not only the global impact of risk from major tsunami hazards, but also the global responsibility to search for means to reduce tsunami risk and enable communities to develop more resilient frameworks and practices for managing such risk. This project represents a substantial step forward in reducing tsunami risk, but more research, development, and practical action are needed for coastal communities to become resilient in practice.

Expanding the Scientific Frontier in Early Tsunami Detection and Warning

In important ways, findings from the activities undertaken in this project have expanded the scientific frontier in early tsunami detection and warning. First, the scientific principle of using the thermocline layer of warm equatorial waters to extend underwater acoustic communication over distances of 25 to 30 kilometers. has been established and documented in chapters 8 and 9. Although this principle had been reported earlier in ocean science and hazards research (Freitag et al. 2010; Comfort et al. 2012), underwater acoustic communication at distances of 25–30 kilometers. in an actual ocean environment had not previously been documented. This finding is especially relevant for Indonesia, given its location on the "Ring of Fire" where the slow movement of Earth's tectonic plates generate earthquakes, tsunamis, and volcanic eruptions as continuing deadly threats. Validating the use of long-range undersea communication to develop early tsunami detection and warning instruments allows Indonesia to strengthen its Indonesia Tsunami and Early Warning System by providing a valid measure of the displacement of ocean water to generate ocean waves before they reach shore. Further, using undersea networks of acoustic communication would greatly reduce the cost of establishing tsunami detection and warning systems for Indonesia with its major coastal cities at risk. Other nations exposed to tsunami risk, such as Japan, use undersea cabled tsunami detection systems (Kanazawa 2013), but such systems would be prohibitively costly for Indonesia, with approximately 81,000 kilometers. (50,300 miles) of coastline.

Second, data confirming a change in water pressure that indicates an emerging tsunami wave can be transmitted via the underwater detection network to a shorestation connected to the satellite system and further transmitted to the Indonesian scientific agency BMKG for confirmation of an imminent tsunami. BMKG can then transmit a confirmed tsunami warning to provincial and municipal emergency operations centers (EOCs) to alert communities at risk. The final step is the communication from the local EOCs to the neighborhood populations to advise and guide evacuation strategies. This "last mile" communication has been the most difficult to achieve, attested by the sobering number of deaths reported in recent tsunamis: Aceh in 2004, Mentawai in 2010, and Palu and Sunda Strait in 2018. The design, modeling, and testing of an electronic neighborhood network reported in chapters 3, 4, 6, and 7 create a novel wireless hot spot network that moves with the members of a neighborhood using a connected smartphone app as they follow evacuation routes, adapted in real time, to safe shelter.

Third, these two networks, undersea and neighborhood, have been designed, tested, implemented in actual neighborhood and ocean environments and have been documented as working and ready to strengthen the two most vulnerable points in InaTEWS: the rapid detection and validation of an incoming tsunami and the warning and interactive communication among neighbors to guide evacuation at the local neighborhood level. Given the extraordinary set of external circumstances in which the development of this prototype system took place, including a global pandemic that disrupted the implementation of the undersea network in 2020 and 2021, not all components of the prototype have worked together at the same time. What has been achieved is a strong, tested, and documented road map for the full development of both networks and their integration into InaTEWS to strengthen the timely, comprehensive system of early detection, warning, and guidance for evacuation for Indonesia.

The final challenge is to build on the concepts, experience, and findings from this prototype development project to implement a fully integrated, working early tsunami detection and warning system that could supplement InaTEWS and be deployed in other coastal cities in Indonesia. Such a system could be extended to other tropical countries that benefit from the thermocline layer in warm equatorial waters. The engineering challenges involved in implementing such a system warrant the investment in building this prototype system to design, test, and validate this approach before formal adoption in a working early tsunami detection and early warning system. The set of exercises and related findings reported in this book provide a sound basis for a full-scale implementation of the undersea and neighborhood networks for early tsunami detection and warning. The reported findings represent a first step in a major advance in the capacity of coastal cities and nations to manage the recurring risk of tsunamis with greater resilience and reduced loss of lives, property, and social disruption.

Advances in Ocean Science

Several advances in the scientific understanding of the ocean environment are already proposed or in progress. First is a proposed international undersea research station or hub that would connect with other sensors to monitor the changing status of the ocean that builds directly on findings reported in this book. This hub would use underwater acoustic communication to share data with other nations regarding the state of the ocean's health and potential indicators for coastal hazards. The status of the oceans is a global problem. Using the concept of an international hub, the design of acoustic networks becomes a possible strategy to share data costs. Remotely operated vehicles (ROVs) can now operate at ocean depths of 2,000 meters.

The developing technology in cable-based ocean observing systems needs to have multiple sensors at one or more specific nodes along the cable system. Several novel technologies are integrated into a one-node cable system, or hub. Japan has developed DONET (Dense Ocean Floor Network System for Earthquakes and Tsunamis) dedicated to observing (near) real-time earthquakes and tsunamis from the seafloor over long periods of time, using a submarine cable network. This program was started in 2011 and began operating in 2015 and is focused on earthquake, tsunami, and geodetic observations. There are seven science nodes, in which every node in the cable array connects to four or five sensors such as pressure gauges or seismometers. Another system implemented in Japan is the S-net (Seafloor Observation Network for Earthquakes and Tsunamis along the Japan Trench) installed in 2018 (Kanazawa 2013; Tanioka and Gusman 2018).

Monterey Accelerated Research System (MARS)

The Monterey Bay Aquarium Research Institute in Moss Landing, California, has developed a nearly similar concept of the cable-based underwater observatory, named Monterey Accelerated Research System (MARS). The system connects a 52-kilometer fiber-optic cable from the shoreline in Monterey Bay extending to the west into the Pacific Ocean. In this single-ended system, the end of cable anchored at 891 meters in the deep sea is the main hub with several dedicated nodes, and these nodes connect to various sensors as needed. Several options of science and observation experiments can be attached to this main hub using either cable connectors directly or a horizontal acoustic system (Freitag et al. 2010).

SMART cable initiative

Worldwide efforts have been developed to use the underwater telecommunication cables crossing the oceans and waters to become the network of real-time data for a disaster mitigation and environment monitoring system. An underwater cable telecommunication system needs optical repeaters approximately every 60 kilometers, thus creating the opportunity to add earthquake, tsunami, or environmental sensors to the repeaters. To bring this concept to the operational stage, three United Nations agencies—the International Telecommunication Union, the World

Meteorological Organization, and the Intergovernmental Oceanographic Commission of UNESCO (IOC-UNESCO)—established a joint task force in 2012 with the aim of incorporating environmental monitoring and tsunami sensors into transoceanic submarine cable systems, hence called the SMART (Science Monitoring and Reliable Telecommunications) cable system. However, no underwater telecommunications monitoring system is in place today.

The concept represents another great idea from the oceanographers' community. The ocean is key to understanding societal threats including climate change, sea level rise, ocean warming, tsunamis, landslides, and earthquakes, but the ocean is difficult and costly to monitor. Integrating sensors into future undersea telecommunications cables or acoustic cables represents the mission of the SMART subsea cable initiative (Howe et al. 2013).

Indonesian cable-based tsunameter program

The most advanced development in the InaTEWS program in 2021–2022 was successfully deployed as the InaCBT (Indonesian cable-based tsunameter) pilot project in Labuan Bajo, Flores Island, in January 2022. This single fiber-optic cable extends from the Labuan Bajo shoreline 54.5 kilometers to the north in the Flores Sea. Two nodes, named LBB-01 and LBB-02, are attached to the cable at 34 and 54.5 kilometers, respectively, with each node consisting of one Paroscientific pressure sensor, a three-axis accelerometer, and temperature and humidity sensors. See figures 12-1 and 12-2.

FIGURE 12-1. Cable route for InaCBT Labuan Bajo.

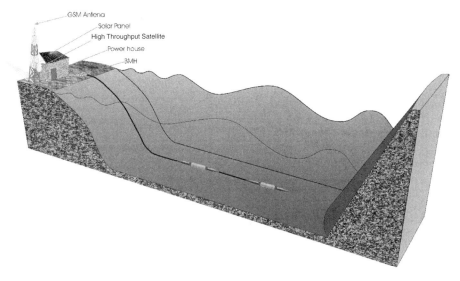

FIGURE 12-2. Concept of Ina CBT system in Labuan Bajo.

The InaCBT system is now performing well, with a sampling rate up to 125 Hz and 1 Hz, respectively, for accelerometer and pressure gauge sensors. This InaCBT pilot project is working well as designed. Several minor earthquakes of magnitude $M < 6.0$ were detected both by the pressure sensors and the accelerometer. In the next step, InaCBT will be integrated with the hybrid Siberut system. The cable from the shoreline to LBB-01 will still exist, but the cable between LBB-01 and LBB-02 will be replaced by a horizontal acoustic link. Figure 12-3 shows the data output from the two sensor nodes.

The challenge for future development is how to integrate the concept of InaCBT and the Siberut hybrid system into one observation system. It is expected that a one-cable system very similar to the InaCBT system will accommodate several sensors. One hub similar to MARS is attached to the offshore end of the cable. The hub may connect to the cabled sensor directly or link acoustically to other remote sensors within a radius of 20 kilometers from the hub.

The InaCBT system has been proven to work well during seismic events, although they were small earthquake events. For example, BMKG reported two consecutive earthquakes shaking the Flores Sea on July 23, 2022, at 13:09:17 WIB M5.4 and 14:35:46 M5.7, respectively. See figure 12-4. These two events were detected on both LBB-01 and LBB-02. Pressure gauge recorders were triggered from normal mode to tsunami mode

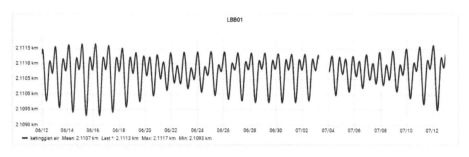

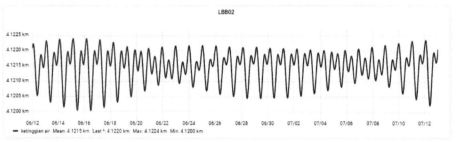

FIGURE 12-3. The ocean bottom units LBB-01 (top) and LBB-02 (bottom) are working well, measuring dynamic pressure since the first week of February 2022 until present. Sampling rate 1 hertz for pressure gauge.

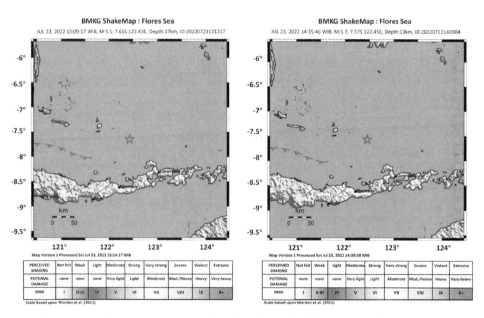

FIGURE 12-4. Two consecutive earthquakes shook the Flores Sea on July 23, 2022, at 13:09:17 WIB M5.4 and 14:35:46 M5.7, respectively (BMKG 2022).

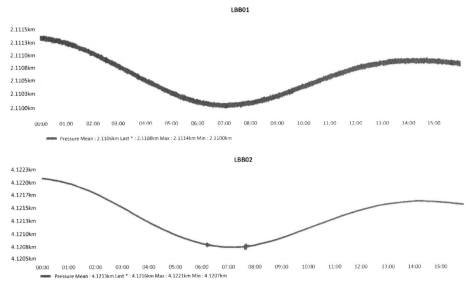

FIGURE 12-5. Sea level anomalies (spikes) detected by pressure gauges in (top) LBB-01 and (bottom) LBB-02 due to an earthquake event in the Flores Sea on July 23, 2022, at 13:09:17 WIB M5.4 and 14:35:46 M5.7, respectively.

after 3-centimeter pressure difference thresholds (i.e., sea level change) for three consecutive 15-second time intervals were confirmed (figure 12-5). The signals of the earthquakes were also recorded by the three-axis accelerometer (figure 12-6). There is no case for a tsunami event at the moment.

These configurations designed for ocean monitoring become more economical, effective, and efficient in terms of cable length, number of sensors, variation of observations, spatial coverage, and system sustainability. Depending on the location and purpose of the deployed system, sensors attached to the system may include pressure gauges, accelerometers, seismometers, upward-looking ADCPs (acoustic doppler current profiler), CT (conductivity and temperature) sensors, hydrophones, underwater cameras, or other specific sensors.

ADCP enables measurement of the three-dimensional current velocity at several depth layers. CT sensors measure salinity and temperature at certain depth layers. Data from ADCP and CT are very important to estimate the water mass transport over a long period of time, allowing oceanographers to monitor climate change and its anomalies, such as the Indonesian Throughflow, El Ninō, La Niña, and the Indian Ocean Dipole.

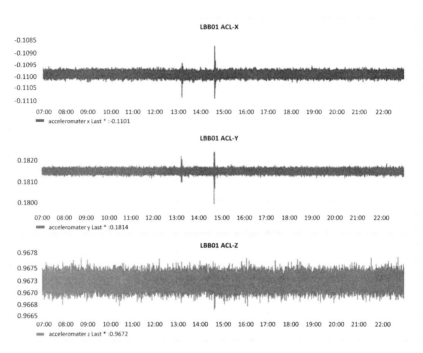

FIGURE 12-6. Spikes detected by three-dimensional accelerometer in (above) LBB-01 and (opposite page, top) LBB-02 due to an earthquake event in the Flores Sea on July 23, 2022, at 13:09:17 WIB M5.4 and 14:35:46 M5.7, respectively. Only *x*- and *y*-axes indicate spikes, whereas *z*-axes have no spikes.

Sensor hydrophones allow the determination and recorded direction of sound that originates from living marine organisms, natural processes such as an earthquake or landslide, and human activities. Scientific sensors can be connected to the hub in two different ways: directly connected through fiber-optic cable or connected remotely using a horizontal acoustic link, with acoustic modems installed both at hub and remote sensors (up to 25 kilometers away). Each instrument or sensor connector can be plugged or unplugged to the hub, using the working-class deep-sea ROV, with two arms, that can operate at ocean depths of 2,000 meters. BRIN is in the process of procuring this deep-sea working-class ROV for 2023. The prototype hybrid system, such as the one deployed near Siberut Island, provides essential knowledge of how acoustically connected sensors to the hub significantly decrease costs by reducing the cable length and enhance efficiency by increasing the number of sensors connected to the hub.

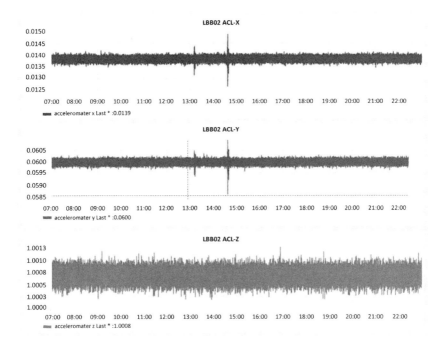

UN Decade of Ocean Science for Sustainable Development

Around 10 percent of the global population, approximately 680 million people, currently live in low-lying coastal areas that are prone to tsunami threat. This total is projected to reach more than one billion by 2050 (IPCC 2019). Since the 1990s, about 295 tsunamis were confirmed and observed worldwide. Some have resulted in devastating loss of life and economic damages, such as the 2004 Indian Ocean tsunami with 230,000 deaths and US$10 billion in economic loss and the 2011 Tohoku tsunami with 18,000 lives lost and US$220 billion in economic damage. The 2009 Samoa tsunami resulted in losses equal to 30 percent of its gross national product, showing that the small island developing states (SIDs) and the least developed countries (LDCs) are more exposed to tsunami risk.

Since 2004, major advances in tsunami warning and mitigation systems around the globe have been achieved under the auspices of IOC-UNESCO, and entirely new warning and mitigation systems have been established in the Indian Ocean, Caribbean, South China Sea, and Northeastern Atlantic Ocean and Mediterranean Sea basins (IOC 2016). Still, critical capability gaps remain. For example, in an extreme near-field tsunami event, a tsunami can strike in as little as five to ten minutes after origin. These gaps leave insufficient lead time for national or local governments to produce public alerts. Any ordered actions are based on a generic and preplanned response with broad assumptions containing high degrees of uncertainty. In addition, accurate inundation forecasts cannot be produced for many locations around the globe due to lack of detailed coastal bathymetry data. Further, for tsunamis generated by poorly understood or nonseismic sources—such as landslides (submarine, subaerial, and combined), submarine volcanic eruptions (pyroclastic flow, caldera collapse, underwater explosion, and blast), or weather-induced tsunamis (meteo tsunami), as well as atmospheric-induced tsunamis that occur in inland waterways or large lakes—virtually no capability to produce forecasts in real time exists (Schindelé 2022). Based on the latest survey, more than 50 percent of tsunami-threatened countries do not have the tsunami evacuation maps and plans necessary to respond quickly and effectively to tsunami warnings.

Under the UN Decade of Ocean Science for Sustainable Development, a framework of action was developed to address these gaps. By implementing transformational gains related to rapid tsunami detection, measurement, and forecasting capability, this action framework will also increase community resilience and capacity to respond to warning systems. To comply with the components of the UN Office for Disaster Risk Reduction's people-centered early warning systems and three pillars of the tsunami warning and mitigation system, these actions will focus on improving *risk knowledge, monitoring and warning, warning dissemination and communication, response capability*, and *capacity development and attention to SIDs and LDCs* through advances in science, observation, and technologies.

Risk knowledge could be improved by enhancing the understanding of the tsunami hazard through strengthening the knowledge of potential tsunami sources and by understanding tsunami impacts to critical infrastructure, marine assets, and the built environment (cities) and how to minimize them. The component of monitoring and warning could be improved through rethinking ocean observation and reducing uncertainty in global tsunami forecasts. The need to detect more quickly and measure tsunamis directly through ocean observation underscores the vital importance of including instrumentation of undersea cables in this process. Critical tsunami generation parameters can be identified through the optimal use and real-time sharing of new and existing sensors and data. There is also a strong need for requiring good resolution of complete bathymetric and topographic data.

Considering the importance of the multihazard early warning framework as a specific target of the Sendai Framework for Disaster Risk Reduction (United Nations 2015), warning dissemination and communication functions should be improved by ensuring the full integration of tsunami services within a multihazard early warning framework. This framework would facilitate the development of warning and communication options that are appropriate to geographic, demographic, and infrastructure conditions for the timely dissemination of warnings.

To improve response capability to tsunami hazards, several activities should be included. Tsunami evacuation maps must be available for all coastal communities to ensure that 100 percent of tsunami-prone communities around the world are prepared and resilient. Plans to minimize impacts to critical infrastructure and marine assets need to be in place to enable quicker post-tsunami restoration of services. Such plans will ensure mainstreaming of tsunami disaster risk-reduction policies in urban and spatial planning at the city and municipality level.

Conclusion

Transformational gains in the tsunami warning and mitigation system will represent a significant component within the Ocean Decade's tsunami program. Specifically, the system would contribute to achieve the goals of the UN Decade of Ocean Science for Sustainable Development, that is, a predictable ocean, a safe ocean, and an accessible ocean (IOC 2020, annex 1). In continuing to expand the frontier of ocean science, findings reported in this book from the Hazard SEES research project and its continuing phases are complementary and strongly contribute to the science we need for the ocean we want as the vision of the UN Decade of Ocean Science for Sustainable Development.

Note

1. The proposal to undertake the Hazard SEES project, funded by the National Science Foundation in 2013, stated the project's goal as to "define, design, and demonstrate an interdisciplinary, dynamic process that will transform societal understanding of risk and enable self-organized, collective action to support the resilient management of hazards." NSF #1331463: "Hazard SEES Type 2: From sensors to tweeters: A sustainable sociotechnical approach for detecting, mitigating, and building resilience to hazards." September 1, 2013–August 31, 2017. Project description, 1.

References

BMKG (Indonesian Agency for Meteorology, Climatology, and Geophysics). 2022. "BMKG Shake-Map: Flores Sea." Jakarta: Government of Indonesia. Processed July 23, 2022.

Comfort, L. K., T. Znati, M. Voortman, Xerandy, and L. E. Freitag. 2012. "Early detection of near-field tsunamis using underwater sensor networks." *Science of Tsunami Hazards* 31, no. 4: 231–43.

Freitag, L., K. Ball, P. Koski, S. Singh, and E. Gallimore. 2010. "Acoustic communications for deep-ocean observatories: Results of initial testing at the MBARI [Monterey Bay Aquarium Research Institute] MARS node." In *OCEANS'10 IEEE SYDNEY*, 1-6. IEEE.

Howe, B. M., B. K Arbic, J. Aucan, C. R. Barnes, N. Bayliff, N. Becker, R. Butler, S. Doyle, S. Elipot, G. C. Johnson, et al. on behalf of the Joint Task Force for SMART Cables. 2019. "SMART cables for observing the global ocean: Science and implementation." *Frontiers in Marine Science* 6, article 424: 1-27. doi:103389/fmars2019.00424.

IOC (Intergovernmental Oceanographic Commission). 2016. *IOC Capacity Development*. Sec. 3.2.5. Tsunami warning and mitigation. Geneva: UNESCO.

IOC 2020. *IOC Circular Letter*, 2815. December 21, 2020.

IPCC. 2019. *IPCC Special Report on The Ocean and Cryosphere in a Changing Climate*, H.-O. Pörtner, D. C. Roberts, V. Masson-Delmotte, P. Zhai, M. Tignor, E. Poloczanska, K. Mintenbeck, et al., eds. Geneva: IPCC. https://www.ipcc.ch/sroc/2019.

Kanazawa, T. 2013. "Japan Trench earthquake and tsunami monitoring network of cable-linked 150 ocean bottom observatories and its impact to earth disaster science." *2013 IEEE International Underwater Technology Symposium (UT)*: 1–5. doi:10.1109/UT.2013.6519911.

Schindelé, F. 2022. Presentation and workshop on "Further Challenges for Warnings of Tsunamis." UN Ocean Decade Safe Ocean Satellite Lab, April 6–7, 2022.

Tanioka, Y., and A. Gusman. 2018. "Near-field tsunami inundation forecast method assimilating ocean bottom pressure data: A synthetic test for the 2011 Tohoku-oki tsunami." *Physics of the Earth and Planetary Interiors* 283 (October): 82–91.

UNESCO. 2021. *UN Decade of Ocean Science for Sustainable Development (2021–2030)*. Geneva, Switzerland: UNESCO. https://www.oceandecade.org/.

United Nations. 2015. *Sendai Framework for Disaster Risk Reduction 2015–30*. Geneva, Switzerland: UN Office of Disaster Risk Reduction.

Afterword

Louise K. Comfort

Many threads are entwined in the tapestry of curiosity, courage, perseverance, and patience, woven together over a period of nine years, that led to the production of this book. The book, in turn, builds on a longer international research collaboration that began in 2004. Looking back, several of these strands run deeper and stronger than others and provide the frame within which the other threads are interwoven. First, the commitment to science and the discovery of new methods of exploring communications in both undersea and neighborhood networks to support early tsunami detection and warning served as the primary motivation for the project, one that was shared by all members of the research and implementation team. This commitment underscored the persistence to overcome unanticipated difficulties, and it only deepened over time. If there was a problem, what did we miss? How could it be done better? What is needed to get this project right? The search continued, time after time.

Second, the profound commitment to humanitarian values bridged the knowledge and skills of the research team with the resources provided by the funding organizations, research institutions, and governmental agencies to enable this project to move forward and continue after encountering multiple unanticipated obstacles. The searing recollection of thousands of lives lost due to lack of tsunami detection and warning kept this complex network of individuals, organizations, and institutions focused on finding a better way. It was a problem that we could not fail to solve.

Third, the promise of integrating new technologies with human ingenuity and innovation enabled members of the research team to devise

313

alternative strategies of achieving the goal of timely, valid communication through both the neighborhood wireless network and the undersea acoustic network when preliminary strategies were found inadequate. Discovering the balance between cost and effectiveness for workable networks, both on land and under the sea, proved a continuing challenge for this project. Determining the requirements for operating technical communication systems in a degraded disaster environment on land and in a rugged deep ocean environment is central to this balance, and this prototype system contributed substantially to that effort.

Fourth, the continual process of learning provided renewed hope and commitment in multiple dimensions. Working across cultures and languages, appreciating differences in administrative procedures and organizations, and recognizing limits of initial plans but adapting them to accomplish the intended goal more realistically stretched the imagination, skills, and endurance of this international, interdisciplinary research team to achieve rare accomplishments. The mobile wireless electronic network connecting with the Iridium satellite to provide a movable wireless communication network at the neighborhood level in a disaster-degraded environment had never been done before. Nor had an undersea network achieved acoustic communication over 25 kilometers in deep equatorial waters before. These achievements, viewed from the larger perspective of early tsunami detection and warning, may be modest, but they represent sound steps of discovery. Developed and integrated with other technologies and organizational systems, these discoveries contribute to building a viable, reliable, working system for early tsunami detection and warning that will save lives.

Finally, the international vision that shaped this project from its beginning has only been sharpened and clarified over the years of this project. Tsunamis are a global threat, and the cost of failing to monitor and detect them falls not just on Indonesia, but on other countries as well. This prototype system has demonstrated key steps toward reducing the risk of tsunamis for coastal cities and shows bold promise for managing this risk on a global scale. The findings presented in this book chart a path toward the full development of early tsunami detection and warning systems and represent a substantive contribution to the United Nations Decade of Ocean Science.

ACKNOWLEDGMENTS

The culmination of ideas, thoughts, contributions, encouragement, and countless acts of support and assistance from many individuals, groups, and institutions made this book possible. The list is too long to name everyone who supported this effort, but several deserve special mention. We warmly thank our colleagues who worked with us in the first years of the Hazard SEES project: Kathleen Carley, Michael Kowalchuk, Sera Linardi, and Daniel Mossè. We acknowledge with thanks and appreciation the National Science Foundation for the financial support that initiated this project and continued its evolution, awarding three grants over a period of twelve years. Two grants supported the early exploratory work for this project under NSF grants, no. 0549119 and no. 07294562. A third NSF grant, OCE-Hazard SEES, no. 1331463, supported the research reported in chapters 1 through 9 and served as the sponsoring partner for funding from the Partners in Enhanced Engagement in Research program of the United States Agency for International Development that also provided funding for the research reported in chapter 2. Findings from research reported in chapters 10 and 12 received funding from the Swiss Re Foundation, Zurich, Switzerland, continuing the research initiated under the Hazard SEES project. The third deployment cruise was also supported by the Yayasan Anak Bangsa Bisa (YABB) Foundation of Indonesia, and the fourth deployment cruise received additional support from Indonesia's Agency for the Assessment and Application of Technology (Badan Pengkajian dan Penerapan Teknologi, BPPT) and its successor agency, National Research and Innovation Agency (Badan Riset dan Inovasi Nasional, BRIN).

Throughout this period, 2005–2022, three research institutions supporting this project—the University of Pittsburgh, the Bandung Institute of Technology, and Andalas University—have contributed time, resources, knowledge, and patience to enable this project to overcome obstacles and continue the investigation. In 2013, the interdisciplinary research team expanded to include the Woods Hole Oceanographic Institution, Carnegie Mellon University, and Northwestern University for the implementation of the research design in Padang and the Mentawai Sea, Indonesia. Importantly, BPPT contributed substantial funding, personnel, equipment, knowledge, and skill to support the completion of this project. We are deeply indebted to Dr. Hammam Riza, former head of BPPT, for his insight, substantial financial resources, and support of the project. We also thank Dr. Rahardjo Sasono and his team from the Technology Center for Disaster Risk Reduction Technology (Pusat Teknologi Reduksi Risiko Bencana, PTRRB) and the Center of Marine Survey Technology (Balai Teknologi Survey Kelautan–Balai Teksurla) for their design, implementation, and integration of the electronics component of the undersea acoustic network. We are also deeply indebted to Professor Dwikorita Karnawati, head of Indonesia's Agency for Meteorology, Climatology, and Geophysics (Badan Meteorologi Klimatologi dan Geofisika, BMKG), for her encouragement and grasp of the scientific innovation that this project provided.

We acknowledge, with thanks and appreciation, the officers and staff at the University of Pittsburgh who supported this project over the years: Jennifer Woodward, director, Office of Research; John Keeler, dean, Graduate School of Public and Institutional Affairs; and Alisha Cuniff, research coordinator and director of finance, Graduate School of Public and Institutional Affairs. Similarly, we acknowledge the administrative officers, staff, and research assistants at our partner institution Bandung Institute of Technology, for their careful financial management and continued support: Dr. Sri Maryati, dean of the School of Architecture, Planning, and Policy Development; Professor Benedictus Kombaitan, former dean of the School of Architecture, Planning, and Policy Development; Dr. Chalid Idham Abdulah, director of the Lembaga Afiliasi Penelitian dan Industri–Institute Technology of Bandung (LAPI ITB) Foundation; and Andi Idham Asman and InIn Wahdini, research assistants. Similarly, we acknowledge the administrative officers and staff at our partner institution, Andalas University for their continued support, especially Professor Tafdil Husni, former rector of Andalas University.

We are especially grateful to Elodie de Warlincourt and the Swiss Re Foundation for their thoughtful guidance and timely and invaluable financial support for the undersea network. We also thank Yudas Sabaggalet, mayor of Mentawai Regency, and Dodi Prawiranegara, police chief of Mentawai Regency, for local support in hosting the shore station for this project. Further, we acknowledge Antoine de Carbonnel of GOJEK Company and the YABB Foundation and staff—Nadiah Hanim, Andre Prasetyo, Monica Oudang, and Dewi Tan—for their strong and vital support of the third deployment. We thank Bambang Haryono and his company, PT Delta Anugerah Bahari Nusantara, for the professional work in implementing the undersea network, as well as Julian S. Khou and the Communications Cable System Indonesia Company in fabricating and testing the cable and canister for the undersea network. We are grateful for the patience and thoughtful guidance of Erin Johnson, editor, and David Miller, president of Island Press, in managing the production of this book.

We also acknowledge, with thanks and appreciation, the dedication and commitment of our colleagues and coauthors of the chapters included in the book. Special thanks are due to Fuli Ai who, in addition to serving as a postdoctoral researcher and coauthor of chapters, also formatted all figures for the book to the required specifications for production.

Most importantly, we thank our families, who offered warm encouragement, thoughtful comments, unwavering love and support, and unfailing patience as we worked through the many external challenges and unanticipated delays in the completion of this manuscript.

Louise Comfort
Oakland, California, USA

Harkunti P. Rahayu
Bandung, Indonesia

LOUISE K. COMFORT is professor emerita and former director of the Center for Disaster Management in the Graduate School of Public and International Affairs at the University of Pittsburgh. She served as principal investigator of the National Science Foundation project, No. 1331463, "Hazards SEES Type 2: From sensors to tweeters: A sustainable sociotechnical approach for detecting, mitigating, and building resilience to hazards," and the supplementary Swiss Re Foundation grants, "Deploying and testing—early Tsunami warning system," November 20, 2018, to December 31, 2019; "Deploy, test, prototype early tsunami detection and warning system" January 1, 2020, to November 31, 2021; and "COVID-19 relief," June–August 2020, which funded this research. She has been a fellow of the National Academy of Public Administration since 2006. Her recent books include *The Dynamics of Risk: Changing Technologies and Collective Action in Seismic Events* (Princeton University Press, 2019) and *Global Risk Management: The Role of Collective Cognition in Response to COVID-19*, coedited with Mary Lee Rhodes of the University of Dublin (Routledge, 2022). Comfort has worked collaboratively with Indonesian colleagues since 2004.

HARKUNTI PERTIWI RAHAYU has been a faculty member of the Urban and Regional Planning Department in the School of Architecture, Planning, and Policy Development at Indonesia's Institute Technology of Bandung for almost thirty years. She earned her PhD from Kochi University of Technology in Japan in 2012. She has served as president of the Indonesian Disaster Expert Association since 2017; chair of Working Group 1 of the UNESCO Intergovernmental Ocean Commission (IOC)/Intergovernmental Coordination Group (ICG) for the Indian Ocean Tsunami

Warning and Mitigation System since 2012; and a member of the ICG Tsunamis and Other Hazards related to Sea Level Warning and Mitigation System (TOWS WG) since 2015. She also currently serves as chair of a task team on tsunami disaster management and preparedness of ICG TOWS WG (2022–2024) and is a member of the Scientific Committee for the UN Ocean Decade Tsunami Program (2022–

2024). She was a member of the National Research Council from 2019 to 2020 and was a recipient of the Newton Prize 2019 Award for her work on developing coastal cities and community resilience. Her work in disaster planning since 2002 has helped improve the capacity of community and local governments at the county and region levels, and she has contributed to guidelines and policy recommendations at national and international levels.

Rahayu has also received international research grants from the National Academy of Sciences USAID PEER Cycle 3 and Cycle 6 (2015–2022), the Newton Fund (2016–2023), and the British National Environment Research Council (2019–2021) for improving coastal city and community resilience toward tsunamis and multihazards, making disaster risk reduction policy recommendations, and converging climate change adaptation and disaster risk reduction strategies into coastal metropolitan/agglomeration planning using sociotechnical approaches for detecting, mitigating, and building resilience to hazards. She was an active member of the Hazard SEES research team from 2013 to 2019 and deployed and tested hybrid tsunami detecting systems from 2018 to 2021.

CONTRIBUTORS

Senior Coauthors

Keenan Ball

KEENAN BALL is a senior engineer in the Department of Applied Ocean Physics and Engineering at Woods Hole Oceanographic Institution, Woods Hole, Massachusetts. He holds a BS in electromechanical engineering from the Wentworth Institute of Technology, 2000, and an MS in electrical engineering from the University of Massachusetts Dartmouth, 2003. His research interests include the use of acoustic arrays to study mammal communications, transducer and acoustic array design and implementation, acoustic communication in autonomous underwater vehicles and design of electronics and mechanical hardware for use with acoustic communications systems.

Abdul Hakam

ABDUL HAKAM is a professor in the Civil Engineering Department at Andalas University in Padang, West Sumatra, Indonesia. He completed his bachelor's degree in civil engineering at Andalas University (1986–1991), where his final project described the structural analysis for reinforced concrete–multistory buildings. He completed his master's program cum laude in geotechnical engineering at the Bandung Institute of Technology in Indonesia (1992–1994). He obtained a PhD from the University of New South Wales in Sydney, Australia, in geotechnical engineering (1997–2001), where his doctoral dissertation described the optimization method for geotechnical structures due to dynamic loads, including earthquake loads. He also takes on many practical activities as a researcher and consultant in the general field of civil engineering, especially geotechnical structures. Besides studying soil mechanics and foundations, in his research and community service he often conducts studies on earthquake-safe houses that are efficient and applicable to ordinary people.

Lee Freitag

LEE FREITAG is a principal engineer in applied ocean physics and engineering at the Woods Hole Oceanographic Institution (WHOI), Woods

Hole, Massachusetts. He holds BS and MS degrees in electrical engineering from the University of Alaska, Fairbanks. He has spent his professional career in the Applied Ocean Physics and Engineering Department at WHOI, where he has worked since 1992. He has worked primarily on underwater acoustic communication and navigation to support ocean science, unmanned undersea vehicles, and Navy submarine systems. In the late 1990s, he worked on the development of long-range acoustic communications for Navy submarine systems, and the methods developed for that project are still in use today. That work was followed by development of the Micromodem, which initially provided command and control of small vehicles performing shallow-water missions. At WHOI, he manages a team of engineers who focus on acoustic communications, navigation, and related system integration. He has participated in more than fifty scientific cruises around the world, including the Arctic Ocean north of Alaska and Svalbard.

Febrin Anas Ismail

FEBRIN ANAS ISMAIL is a professor of civil engineering and director of the Center for Disaster Mitigation, Andalas University, Padang, Indonesia. He received his bachelor and master of engineering degrees from the Bandung Institute of Technology in Indonesia and his PhD in engineering from Yokohama National University in Japan. His research interests are in the design and construction of buildings and structures to withstand damage from earthquakes and tsunamis. He served on international reconnaissance teams following the 2004 Indian Ocean tsunami and the 2009 Padang earthquake as an engineering expert in seismic-resistant construction. He is a member of the Indonesian Association of Disaster Experts, as well as international professional associations in engineering. Ismail has published articles in international journals, including *Natural Hazards* and *Journal of Earthquakes and Tsunamis*, and has organized several international conferences on earthquake hazards and tsunami risk. He has served on the international and interdisciplinary team Hazard SEES to develop and test an early tsunami detection and warning system, funded by the National Science Foundation, 2013–2018, and the subsequent extensions funded by the Swiss Re Foundation and the Yayasan Anak Bangsa Bisa Foundation.

Peter Koski

PETER KOSKI is a research engineer in the Department of Applied Ocean Physics and Engineering at Woods Hole Oceanographic Institution,

Woods Hole, Massachusetts. He earned a BS in electrical engineering from Northeastern University in 1999. His research interests are in the development and application of embedded signal processing technology for underwater acoustic communications, electronic hardware/software design for long-term reliability, real-time signal processing techniques, and data telemetry for long-life remote moored observatories.

Emile A. Okal

EMILE A. OKAL is professor emeritus of earth and planetary sciences at Northwestern University, Evanston, Illinois. He is an expert in the theoretical and observational investigation of tsunamis, with main emphases on the physics of the generation of tsunamis by earthquakes and other sources, and on the reconstruction of historical tsunamis from eyewitness accounts. Okal has led, or participated in, thirty-one post-tsunami field surveys in the Pacific and Indian Oceans and in the Mediterranean Sea. His other professional interests include the origin of deep earthquakes and the preservation of seismic archives. He is an alumnus of the École Normale Supérieure in Paris and holds a PhD from the California Institute of Technology. He is a member of the Seismological Society of America and the European Geosciences Union. He is the author of more than 250 refereed publications and was the 2013 recipient of the European Geosciences Union's Sergey Soloviev Medal. Okal served as the seismologist on the interdisciplinary Hazard SEES team from 2013 to 2019.

Wahyu W. Pandoe

WAHYU W. PANDOE is a principal engineer at the National Research and Innovation Agency of the Republic of Indonesia. From 2017 to 2021, he served as the deputy chairman of BPPT (Indonesian Agency for the Assessment and Application of Technology) for design and engineering industrial technology. He is an expert in the observation and early warning of tsunamis, with main emphases in the platform design and engineering for tsunami observation. His expertise in ocean engineering led him to be appointed as the program director for the operation and maintenance of the Indonesian Tsunami Early Warning System tsunami buoys from 2010 to 2013. Pandoe obtained his bachelor's degree from the Bandung Institute of Technology in Indonesia, majoring in geodetic engineering. He obtained a master's degree in oceanography from the School of Geosciences at Texas A&M University, College Station in December 2000 and a PhD in ocean engineering, specializing in numerical modeling of coastal process and sediment transport, from Texas A&M's School of Civil Engineering in 2004. For his achievements in several areas of public service in Indonesia, Pandoe received five medals from Indonesian presidents, including the Wirakarya Medal in 2020 and the Development Medal in 2008.

James Partan

JAMES PARTAN is a research engineer in the Department of Applied Ocean Physics and Engineering at Woods Hole Oceanographic Institution, Woods Hole, Massachusetts. He earned a BA in astrophysics from Williams College in 1994, a BA in mathematics from the University of Cambridge in 1996, an MS in electrical engineering from the Massachusetts Institute of Technology in 2000, and a PhD in computer science from the University of Massachusetts Amherst in 2013. His research interests include underwater acoustic communication, navigation, and networks; marine mammal acoustics and behavioral sensing; autonomous vehicles; Arctic acoustics and systems; software engineering for real-time embedded systems; electronics hardware design; and signal processing.

Sandipa Singh

SANDIPA SINGH is a senior engineer in the Department of Applied Ocean Physics and Engineering, Woods Hole Oceanographic Institution, Woods Hole, Massachusetts. She obtained a BS in electrical and computer engineering from George Mason University in 1989 and an MS in electrical engineering from the University of Virginia in 1991. Her research interests are underwater acoustic communications; multiuser, multiaccess systems; spread spectrum communications techniques; and error correction coding.

Iyan Turyana

IYAN TURYANA served as a lead instrument engineer for ocean applications at BPPT (Indonesian Agency for the Assessment and Application of Technology) since 1998. Since 2007, he has been a key expert in building a buoy-based tsunami detection system in Indonesia. He received the 2021 Wirakarya Medal for this achievement and has received three other medals. Turyana received his master of engineering degree in electrical and electronic engineering from the Bandung Institute of Technology in Indonesia and an advanced degree in electrical and electronic engineering from Universitè Joseph Fourier (Grenoble I), France. He served as the lead ocean instrument engineer for BPPT on the international and interdisciplinary research team, Hazard SEES, that has been engaged in the design, development, and deployment of an early tsunami detection and warning system in the Mentawai Basin, Indonesia. This research was funded by the National Science Foundation, 2013–2018, and supplemental grants from the Swiss Re Foundation, "Deploying and testing—early tsunami warning system," December 2019–November 2021, to cover the Indonesian costs of deployment. Currently, Turyana is serving as an instrument engineer at the Indonesian Research Center for Geological-Source-Base Disaster, National Research and Innovation Agency (BRIN).

Taieb Znati

TAIEB ZNATI is professor emeritus and former chair of the Department of Computer Science, School of Science and Computing, University of Pittsburgh. He has served as coprincipal investigator for the international and interdisciplinary Hazard SEES project, 2013–2019, funded by the National Science Foundation. In this capacity, Znati directed the design and development of the electronic network at the neighborhood level in disaster-degraded environments. He held joint appointments in telecommunications and computer engineering in the School of Engineering. From 2005 to 2007, Znati served as division director, Computer Networks and Systems, Computer and Information Science and Engineering National Science Foundation. He has also served as director of the CyberSphere Research Initiative and as codirector of the Energy Grid Institute, a public–private partnership focused on energy led by the University of Pittsburgh's Department of Computer Science, Grid@Pitt. His research interests focus on designing fault-tolerant networks, ad hoc networks, and distributed systems in local environments. He has coauthored more than 70 articles published in refereed journals and more than 160 papers in refereed conference proceedings.

Junior Coauthors

Fuli Ai

Since 2019, FULI AI has served as a senior technical director in China Life Property and Casualty Insurance Company, Limited in Beijing. He received his PhD in natural disaster science from Beijing Normal University in 2014. He is an expert in geographic information systems and computer science and has more than twenty years' experience in software engineering and database management. His research field is marine disaster management, especially typhoon and earthquake-tsunami disasters in Asia. He served as a postdoctoral fellow at the Center for Disaster Management, University of Pittsburgh from 2014 to 2018 and as a member of the Hazard SEES international and interdisciplinary research team, supported by the National Science Foundation. As a team member, Ai engaged in designing and developing a community resilience framework based on complex adaptive systems of systems and bowtie architecture theories. The goal of this framework is to provide real-time, reliable, socially aware communication among local governments and local neighborhood organizations in communities exposed to risk during actual emergency events. Ai is a member of the Association for Public Policy Analysis and Management and the Society for Risk Analysis–China, as well as a reviewer of international journals, including *Safety Science, Disaster Risk Reduction, Natural Hazards Review*, and *Environmental Research, Risk Analysis, and Crisis Response.*

Mark W. Dunn

MARK DUNN is assistant dean for academic technology in the School of Law at Duquesne University in Pittsburgh, Pennsylvania. He worked as a research team member with the Center for Disaster Management at the Graduate School of Public and International Affairs, University of Pittsburgh, in various capacities from 1996 to 2018. He teaches undergraduate and graduate courses in database management; advanced database management; data for decision making; data structures; and data mining, data analytics, and big data at La Roche University and Duquesne University. He has a master of public management degree from Carnegie Mellon University and is completing his doctoral studies at the Duquesne University School of Education. His focus is on analyzing large sets of educational data using machine learning methods.

Dennis Giaya

DENNIS GIAYA is an engineer II in the Department of Applied Ocean Physics and Engineering, Woods Hole Oceanographic Institution (WHOI), Woods Hole, Massachusetts. He develops software for embedded digital signal processing, Linux, and low-power microcontrollers to support data communications and sensor telemetry systems. He is a member of the WHOI Acoustic Communications Group.

Kayleah Griffen

KAYLEAH GRIFFEN served as an electrical engineer and data engineer at Woods Hole Oceanographic Institution (WHOI) from 2018 to April

2022. She developed software for sensor systems and communications within the WHOI Acoustic Communications Group.

Echhit Joshi

ECHHIT JOSHI received his master's degree in information sciences from the School of Computing and Information at the University of Pittsburgh in 2019. During his program, he worked as a research team member for the Hazard SEES project. He also interned in a research assistant position for MoSHI (Mobile Sensing + Health Institute), where he analyzed sleep and mobile sensor data to create predictive models for behavioral changes for better health. He is currently a data scientist at Afiniti, where he manages statistical models to improve customer experience.

Yoon Ah Shin

YOON AH SHIN received her PhD in public and international affairs from the Graduate School of Public and International Affairs at the University of Pittsburgh in December 2019. During her doctoral studies, she served as a graduate student researcher in the Center for Disaster Management and as a member of the interdisciplinary research team for the Hazard SEES project funded by the National Science Foundation. She engaged fully in the effort to design interorganizational communication pathways and to build a system dynamics model evaluating the effectiveness of the communication pathways. After graduation, she received two research fellowships: a Socio-Environmental Immersion Postdoctoral Fellowship at the National Social-Environmental Synthesis Center, funded by the National Science Foundation through the University of Maryland–College Park; and a New Carbon Economy Consortium (NCEC) Social Science Fellowship, funded by the Alfred P. Sloan Foundation through Arizona State University's Global Futures Laboratory. She has continued her research interests in interdisciplinary projects, including an interorganizational collaboration system in response to crisis and low-carbon governance to mitigate risk of climate change. Her research integrates social and environmental perspectives to capture and evaluate dynamic interactions between social and environmental systems, focusing on interorganizational coordination and collaboration in response to fast-changing, complex situations.

X. Xerandy

X. XERANDY received his bachelor and master's degrees in electrical engineering from the Bandung Institute of Technology in Indonesia. He joined the Indonesian Agency for the Assessment and Application of Technology (BPPT) in the Marine Survey Technology Department as an instrument engineer in 2008. He enrolled as a PhD student in the Telecommunication and Networking Department in the School of Computing and Information at the University of Pittsburgh in 2011. In 2019, he returned to BPPT to continue his field research. In 2020 to 2021, he assisted with technical supervision and monitoring for the development and deployment of the cable-based tsunami detection program in Indonesia. During his doctoral program at the University of Pittsburgh, he actively participated in the National Science Foundation–funded research project Hazard SEES to design, develop, and test a sustainable sociotechnical approach for community resilience to hazards, and he continued his participation through the subsequent stages of the project in the deployment of the undersea network. His focus is on leveraging sensor and networking technology for undersea and land-based environments that are dedicated to tsunami disaster mitigation. Currently, Xerandy serves as a researcher in electronic sensing, navigation, and network protocol at the Indonesian Research Center for Electronics, National Research and Innovation Agency (BRIN).

INDEX

Page numbers followed by "f" and "t" indicate figures and tables.